IS PLUTO A PLANET?

IS PLUTO A PLANET?

A Historical Journey through the Solar System

DAVID A. WEINTRAUB

PRINCETON UNIVERSITY PRESS

PRINCETON AND OXFORD

Copyright © 2007 by Princeton University Press
Published by Princeton University Press, 41 William Street,
Princeton, New Jersey 08540
In the United Kingdom: Princeton University Press, 3 Market Place,
Woodstock, Oxfordshire OX20 1SY

All Rights Reserved

ISBN-13: 978-0-691-12348-6
ISBN-10: 0-691-12348-9

Library of Congress Control Number: 2006929630

British Library Cataloging-in-Publication Data is available

This book has been composed in Goudy and Raleigh Display

Printed on acid-free paper. ∞

pup.princeton.edu

Printed in the United States of America

5   7   9   10   8   6   4

○

To Carie Lee, MaryAlison, Sarah Beth, and Isaac
and to my parents,
Stanley and Rodelle Weintraub

Contents

Preface

I have taught an introductory course on the astronomy of the solar system to undergraduates at Vanderbilt University for fourteen years. At the beginning of every semester, I ask my students to walk in the shoes of modern astronomers and ask their own questions about the solar system so that we might try to answer them over the course of our semester together in class. Over the years some new questions emerge while others fade into the past, but without fail my students ask, "Is Pluto a planet?"

Is Pluto a planet? is undoubtedly one of the very best and most exciting topical questions in modern astronomy: the story told in this book is an attempt to answer this question in a thoughtful, historically meaningful way that takes readers deep into the history of science and describes many exciting modern discoveries that have led astronomers to the point of asking and trying to answer this question. Most of our technical knowledge about Pluto is not necessary in order to ask and answer the questions that motivate this story; however, readers who wish to learn more of the technical details astronomers know about the object we call Pluto will find a great deal of that information in the appendix.

I am grateful to my students whose enthusiasm and thirst for learning is both inspiring and encouraging, to my friend and artist Adrienne Outlaw whose beautiful and imaginative illustrations grace several pages in this book, to the many professional and amateur astronomers who have graciously shared their work and ideas with me and permitted me to include these in this book, to my parents who have encouraged my efforts to add a book to the family tree, and to Carie Lee and our children, MaryAlison, Sarah Beth, and Isaac, all of whom love books and reading and me enough to have put up with me while I wrote this one.

David A. Weintraub

IS PLUTO A PLANET?

1
What Is a Planet?

O

*I*s *Pluto a planet?* This question appears so simple—clearly the answer is either yes or no—yet the simplicity is misleading. Logically, we must know what a planet is and determine whether Pluto fits those criteria if we are to construct a well-reasoned answer.

The question *Is Pluto a planet?* has stirred the passions of professional astronomers since this enigmatic object was discovered in 1930. In order to understand why this question vexes the professionals, we will first follow the path of early intellectual discovery along which scientists came to recognize that the Earth is a planet. Then we will walk the historical path that led to the discovery of much of the solar system, including the planets Uranus and Neptune, the asteroid belt, and Pluto. Finally, we will focus our attention on important astrophysical discoveries since the discovery of Pluto that have culminated in widespread, popular confusion and impassioned professional debate over the status of Pluto.

At the end of the twentieth century, the debate over the answer to the question *Is Pluto a planet?* spilled into the public domain because several new discoveries—large objects in the Kuiper Belt in our solar system, giant planets orbiting other stars, possible planet-sized objects floating freely through space—all provide new and important ways to think about how to answer this question. We will look at these new discoveries, especially at their impact on how we think about planets and planetary systems, including, of course, how we think about our own solar system.

1

The question *Is Pluto a planet?* illustrates a difficult challenge common to all areas of research and thought: how do we draw the lines we use to categorize objects and ideas? Categorization is one of the first steps in learning: we organize information by similarities and differences. We know that mammals give birth to live young; yet, a platypus lays eggs and is a mammal. Despite its name, a koala bear is a marsupial, not a bear. So which similarities are most important and fundamentally determine membership in a group or class? Which differences are incidental? As is often said, the devil is in the details.

Astronomers like to joke that knowledge of one object—a bright point of light in the nighttime sky—defines a class of objects: stars. The discovery of a second bright point of light that is not identical to the first forces us to create two distinct classes of objects—for example, red stars and blue stars. In this example, the two stars share one important characteristic—they are both bright points of light in the sky—but differ in the apparently important characteristic of color. If we then discover a third object in the sky sharing the original characteristic—a bright point of light in the nighttime sky—but differing in the second characteristic—this third object is yellow—do we conclude that these three objects belong to three distinct categories of objects, with each group being represented by only one example? Or, might we realize that we have discovered three similar objects that differ only in the incidental quality of color? Which is more important, recognizing the similarities among the objects (they are all stars) or emphasizing the differences (colors) between them? At what point should we discard the categories we are using and start anew?

By defining finer and finer categories, we learn about the physical universe in which we live. We use our knowledge both to expand our understanding (Wow, three kinds of stars exist!) and to delimit what we do not understand (Why do stars have different colors?). The problem for us lies in how we define a group (for example, planets) when we know very little about the individual objects in the group and almost nothing about the processes that made or make similar objects and have very few examples of objects that presumably are members of the group.

We cannot answer our question *Is Pluto a planet?* unless we are able to determine the qualities that define the boundaries of the category *planet*. Once we agree on how to define *planet*, we can ask whether a particular object, in this case Pluto, satisfies our criteria. Since we need to define *planet*, we could turn to a dictionary.

The *Oxford English Dictionary* (*OED*)[1] defines the modern word *planet* as deriving from the Old French *planete* out of the Latin *planeta*; in turn, the Latin is derived from the Greek word for wandering star, *planetos*, which in turn evolved from *planasthai*, the verb "to wander." Clearly, we need to understand what the ancient Greeks meant by their word, and the *OED* tells us, giving the *Old Astronomy* (i.e., archaic and no longer used) usage:

> A heavenly body distinguished from the fixed stars by having an apparent motion of its own among them; each planet, according to the Ptolemaic system, being carried round the Earth by the rotation of the particular sphere or orb in which it was placed. The seven planets, in the order of their accepted distance from the Earth, were the Moon, Mercury, Venus, the Sun, Mars, Jupiter, and Saturn.

If you are reading carefully, you will have noticed that according to the ancient Greeks, *the Sun was a planet* but *the Earth was not!* Certainly, no person in the twenty-first century thinks of the Sun as a planet. We all were taught that the Sun is a star, not a planet.

So did the Sun change from a planet into a star? Of course not. Apparently, though, *our understanding* of what is meant by *planet* changed; hence, at the moment when our understanding changed, we reclassified the Sun as a star and the Earth as a planet and discarded the Old Astronomy usage.

Next, we find the *Modern Astronomy* definition:

> The name given to each of the heavenly bodies that revolve in approximately circular orbits round the Sun (*primary planets*), and to those that revolve round these (*secondary planets or satellites*). The primary planets comprise the *major planets*, of which nine are known,

viz., in order of distance from the Sun, Mercury, Venus, the Earth, Mars, Jupiter, Saturn, Uranus, Neptune, and Pluto, and the *minor planets* or asteroids, the orbits of which lie between those of Mars and Jupiter.

Another *OED* definition, given in the *New Shorter OED*, is similar:

Any of various rocky or gaseous bodies that revolve in elliptical orbits about the Sun and are visible by its reflective light, esp. each of the nine major planets (see below); any of various smaller bodies that revolve around these; a similar body revolving around another star.

The "see below" points to a list of *primary planets*, a definition of *minor planet* as an asteroid, and a definition of *secondary planet* as "a planet that orbits another planet, a satellite, a moon."

Whoa! The Moon revolves around the Earth; therefore, the Moon must be considered a *secondary planet*? Certainly, a secondary planet is some kind of planet just as a sweet potato is some kind of potato. Do we normally think of our Moon as a planet? No.

Mars has two potato-shaped moons, Phobos and Deimos, each no larger than a small city, both of which also match this definition. Are objects like Phobos and Deimos, with diameters of only a few miles, planets? No.

The Hubble Space Telescope most definitely is a satellite revolving around a primary planet, the Earth. Is the manufactured Hubble Space Telescope, which astronomers would consider to be a "rocky body," a *heavenly body*? If not, what if NASA hauled a ten-ton boulder into space and launched that boulder into a terrestrial orbit? The boulder is certainly a naturally made rocky body that would be in orbit around a primary planet. What meaningful difference would permit us to distinguish between these two orbiting objects, or would both qualify as *secondary planets*?

According to the latter two definitions, all of these objects might be planets, as are the asteroids in the asteroid belt, even the ones that are smaller than a house or car, since most of these objects revolve in approximately circular orbits around the Sun. However, according to the Modern Astronomy definition, a Sun-orbiting asteroid whose orbit keeps it in between the orbits of Earth and Mars, rather than in between the orbits of Mars and Jupiter, would not be a planet. Similarly, an asteroid in a moderately elliptical orbit between Mars and Jupiter, traveling from an outermost distance just outside of Jupiter's orbit to an innermost distance just inside of Mars's orbit, also would be disqualified as a planet. The *New Shorter OED* definition would allow asteroids with more elliptical orbits and those outside the Mars and Jupiter boundaries to be secondary planets, but why are such objects considered planets at all?

When I read these definitions, I want to know: Who are these Old and Modern astronomers? Am I one? Surely, the Old astronomers were not to be trusted, as they included the Moon and Sun as planets. And what of these Modern astronomers whose definition appears to include objects big and small, natural and manufactured, but only if those objects are in nearly circular orbits and in preferred locations? If either of the modern definitions is correct, far more than nine planets orbit the Sun. In fact, there must be hundreds of thousands, perhaps even millions, of planets in our solar system. Clearly, the commonly accepted notion that nine planets orbit the Sun does not match the *OED* version of the universe.

I, for one, am very dissatisfied when I read the *OED* definitions. I hope you are similarly bothered. Because the dictionary definition of *planet* is essentially worthless, scientifically, we have to probe much deeper into astronomy in order to answer what initially appeared to be a simple question. That is what we will do in the rest of this book.

2
Seven Perfect Planets Made of Aether

O

What would you learn if you could look up and observe the sky, night after night, year after year, unencumbered by the accumulated intellectual burdens and prejudices of the twenty-first-century world? The first thing you would notice is that, like clockwork, the Sun rises every morning and sets every evening. The exact moments and directions of sunrise and sunset change in cyclical patterns that mark the seasons; yet, no matter the season, the Sun appears to sweep out one great circle, moving from east to west in the sky, from exact noon one day to exact noon the next day, in exactly twenty-four hours.

Ancient peoples everywhere must have watched the Sun. By doing so, they learned how to predict the onset of seasonal changes based on the height of the Sun above the southern horizon at noon, the position of the Sun along the eastern horizon at sunrise or the western horizon at sunset, or the lengths and directions of shadows cast by the Sun at different times of day. Those peoples with more advanced skills in measuring time—times of day and times of year—were more likely to prosper and survive, as their ability to stay warm and dry and to find, store, and secure adequate food supplies depended on their expertise in predicting seasonal changes.

Certainly, in times long before written records were kept, our ancestors also must have recognized that the points of light that illuminate the nighttime sky rise and set, just like the Sun. They also would have noticed that different stars rise and set at different times of the year and are organized in recognizable patterns. Further-more, they would have discovered that these stars, as they wheel around the Earth from east to west, never change positions with respect to each other. Their positions seem to be fixed; the stars themselves seem to be attached to an unimaginably large and distant celestial sphere, with the Earth at the center, which makes one com-plete revolution in twenty-four hours.

At first glance, the Sun seems to be attached to the celestial sphere, as the Sun appears to be carried around the Earth from east to west with the stars, once per day. Yet, the most attentive of the ancient sky watchers would have discovered that the stars appear to go around the Earth just a little more quickly than the Sun. On our modern clocks, we would note that the time for a complete cycle of the stars is almost exactly 23 hours, 56 minutes, and 4 sec-onds, nearly 4 minutes less than the 24 hours required for the passage of the Sun from the position in the sky we refer to as noon to its next noontime appearance. Thus, we might imagine that the Sun has two motions. One motion takes the Sun completely around the heavens from east to west in 23 hours, 56 minutes, and 4 sec-onds; the second motion takes it much more slowly in the opposite direction, from west to east, by the equivalent of almost 4 minutes each day.

Since 24 hours is 1,440 minutes, and 1,440 minutes divided by 4 minutes is 360, and since the circumference of a circle has 360 degrees, the Sun moves about 1 degree per day through the fixed stars. At the end of one year, or very nearly 365 days, not only is the Sun back to its starting point among the patterns of the stars, the seasons have also completed one cycle. Thus, the ancient astron-omers would have discovered that they could track the seasons by watching the stars, which for some peoples may have been easier than watching the Sun. The warmth of the Sun obviously influences the weather, the growth of plant life, and the regulation of the sea-sons, whereas the stars do not obviously cause any of these effects; yet, one can easily understand how ancient peoples would have as-sumed that the stars, like the Sun, somehow must affect life on

7

Earth. This primitive logic is ultimately the origin of the practice of astrology and many ancient religious practices.

Once ancient peoples had associated the ability to predict the onset of the changing seasons with the nighttime sky, watching and observing the patterns of the heavens and knowing how to interpret these observations would have become one of the most important jobs within all premodern societies. The Inca, for example, began their calendar year and agricultural year with the first annual predawn appearance of the Pleiades star cluster (in the modern month of June). Furthermore, if the Sun and stars appear to guide and control the day and the seasons, might other celestial objects that are neither Sun nor star exist that control other terrestrial phenomena, or even our lives?

All ancient observers would have noticed the presence and changing appearance of the Moon. Indeed, the 29.5-day period for the phases of the Moon was one of the easiest astronomical periods to quantify, for observers in virtually all ancient cultures. Peoples living along seacoasts would, early on, have associated the Moon with the tides. Now we have a second celestial object of great importance. The Moon, though it operates on a different schedule than the Sun, rising about 50 minutes later every day while paying no attention to the seasons, also rises and sets every day and is visible at night when it is up at night. Whereas the Sun moves through the fixed pattern of stars in 365 days, the Moon zips along a path through the stars and returns to its starting point in just over 27 days. The ancient nighttime observers would have noticed that the path of the Moon, as it travels through the stars, is very similar, though not identical, to the path of the Sun.

The circular path of the Moon, when drawn on the inside surface of our imaginary celestial sphere, intersects the circular path of the Sun at two points. At one of these points, the Moon can pass in front of, and thereby eclipse, the Sun. When this happens, we experience a solar eclipse. At the other intersection point, the Moon can disappear into the Earth's shadow and thereby be eclipsed by the Earth; at such times, we experience a lunar eclipse. Over a period of about eighteen years, these two points of intersection move synchronously around the full circular path of the Sun; that is, they occur at different times of the year. All the points on the celestial sphere at which solar and lunar eclipses can occur fall along the line followed by the Sun. Thus, the thin line along which the Sun travels was given the name *ecliptic*.

Imagine painting a stripe eighteen degrees wide around the inside of our celestial sphere, with the line of the ecliptic running directly through the middle of this band. This band would be wide enough to include the apparent circular path of the Moon, with the Moon's path clipping the top and bottom of the stripe at opposite ends of its orbit. Clearly, the stars in this band are the most important stars for tracking the positions of the Sun and Moon and predicting seasons, and so the ancients quickly identified easily recognizable groupings of stars along this band and gave names to these groupings. On opposite sides of the world, Mayan and Babylonian astronomers independently organized and named these groupings mostly for animals. The Maya included a rattlesnake, a turtle, three birds, a frog, a peccary, a scorpion, a fish-snake, a bat, a skeleton, and an ocelot, while the Babylonians chose a ram, a bull, twins, a crab, a lion, a virgin, a balance, a scorpion, an archer, a goat, a water bearer, and a fish. The Greeks came to call this band in the sky inhabited by these creatures the zodiac and the imaginary figures in the sky the constellations of the zodiac.

While ancient observers were learning about the motions of the Sun and Moon and stars, they could not help but notice that the few brightest points of light in the sky behaved in ways unlike all the other stars. These unusual celestial objects looked like stars, yet were different. Unlike the stars, they changed in brightness with time, becoming fainter and brighter in regular and measurable lengths of time; also, in contrast to the behavior of the stars, they did not twinkle in the night sky and were regularly observed moving backward. Finally, like the Sun and Moon, three of these bright objects moved along a full circular path through the stars within the zodiac and almost, but not perfectly, along the ecliptic. The other two bright objects moved back and forth nearly along the ecliptic, traveling from one side of the Sun to the other, while all the time

Figure 2.1: The constellation *Ursa Major*, the Great Bear, which is found in the northern sky, includes the *Big Dipper*. (Artwork by Adrienne Outlaw ©; photograph courtesy of J. B. Kaler)

moving forward around the ecliptic together with the Sun. Each of these wandering stars circled through the fixed stars in different periods of time, the latter two with periods of approximately 88 and 225 days, the former three with periods of 687, 4,333, and 10,756 days. These five small wanderers, like the stars, did not appear to have any direct influence on the lives of humans, at least in the obvious ways in which the Sun affects us on a daily basis. However, unlike all the other celestial objects, these five wandering stars shared many behaviors with the Sun and Moon and consequently merited and received special attention and special names.

We know these five as Mercury, Venus, Mars, Jupiter, and Saturn. Together with the Sun and Moon, these celestial objects be-

Figure 2.2: Four planets nearly aligned with the ecliptic, as seen at sunset from Maui. The planets are directly above the setting sun and mark out the shape of a hockey stick. Mercury lies closest to the horizon, immersed in the clouds and fading sunset. To the left and above Mercury is Jupiter. Above Jupiter is brilliant Venus, with Saturn shining in the dark, well above Venus. (Picture courtesy of Rick Scott)

came known as the wandering stars, the seven planets. The ancients lacked a logical basis for understanding why some stars wander while others are fixed, but had a means of distinguishing planets from all other objects in the universe. Now, for the first time, we have a working definition of a planet: *wandering star.*

GREAT OBSERVERS AND GREAT THINKERS

From historical records, we know that some ancient peoples were extremely careful and highly skilled observers of celestial patterns and phenomena. They kept records spanning generations, enabling them not only to notice but also to accurately measure cyclical phenomena of great lengths of time. From the Venus tablets from ancient Babylonia and the Dresden Codex of the Maya, we know that both the Babylonians and the Mayan star watchers discovered the 2,920-day period for the cyclical appearance and disappearance of Venus from the morning sky; later, astronomers in both cultures independently discovered a 6,585 ⅓ day cycle for eclipses and determined that the length of the year based on the positions of the stars (the *sidereal* year) was 20 minutes longer than the year based on the position of the Sun (the *tropical* or *seasonal* year).

Whereas the Babylonian astronomers were preoccupied with answering the questions *where* and *when* about observable celestial events, especially the changing positions of the seven planets with respect to the positions of the fixed stars, the pre-Socratic Greek philosophers invested their energies in asking and answering the questions *what* and *why*. Over a three-hundred-year period from 600 to 300 B.C.E., this new way of thinking about the world led to a worldview that calcified as the geocentric, or Earth-centered, universe of Aristotle. Because Aristotle's geocentric universe is fundamental to our understanding of planets, it is worth some effort to investigate the kind of thinking that culminated in Aristotle's ideas for the geometry of the universe.

Thales (624–547 B.C.E.), who introduced geometry to the Greek world and according to legend ended a war by predicting a solar eclipse in 584 B.C.E., began this new way of thinking by asserting that the basic substance of the universe was water. As for the Earth, Thales claimed that it was a flat disk floating in water. Other than the omnipresence of water surrounding the island inhabitants of the ancient Greek world, Thales' universe had no connection to anything observable about the physical universe; however, his ideas regarding thinking about what the universe is made of and what form the universe takes took root. Thales' student Anaximander (610–547 B.C.E.) argued that, rather than water, the world consisted of opposites (for example, hot and cold, justice and injustice) and that these opposites came into being out of the single, never-seen substance known as aperion. Anaximander's student Anaximenes (585–502 B.C.E.) preferred air as the single universal substance. However, Heraclitus (535–475 B.C.E.) opted for fire; Xenophanes (570–474 B.C.E.) chose both earth and water; and Empedocles (484–424 B.C.E.) hedged his bets with four basic substances, earth, air, fire, and water, and two basic forces, love (attraction) and strife (separation). As for structure, Anaximenes said the sky is a revolving crystalline vault, in which the stars are nail heads that reflect sunlight and are pushed in circular paths round the Earth by the denser air. For Xenophanes, the stars were burning clouds, extinguished by day and lit again each night, and the Sun was a collection of fiery particles born anew each morning. None of these pre-Socratic philosophers made any effort to link their ideas about the cosmos to the quantitative observables—the actual positions and brightnesses and the exact times and dates and locations for the rising and setting of the stars and planets—in the heavens. The change from qualitative to descriptive natural philosophy, in which the description had to bear some reasonable resemblance to the perceived physical world, would begin with Aristotle.

Aristotle

Aristotle (384–322 B.C.E.), who wedded his ethical, political, physical, and astronomical philosophies through the concept of *telos*, or purpose, took the reductionism of his predecessors one step further:

he added a fifth element, *aether*. The existence of aether is not an arbitrary idea; rather, Aristotle's ideas about physics required the existence of a fifth element, or quintessence, different from all the rest, with the aether being an eternal, unalterable, unchangeable substance that exists only in the heavens, always moving in perfect circular motion. Because the aether never changed in any way, according to Aristotle, it illustrated how a substance in the universe can have a purpose and can fulfill that purpose.

Aristotle argued that all that happens in the physical universe is governed by the laws of motion. Logically, according to Aristotle, the *natural* direction of motion of earth and water is down, whereas the natural direction of motion of fire and air is up. Sideways motion also is possible if an object is impelled to move forward or backward, but all natural motion is either up or down. In addition, all up and down and sideways motions ultimately end; all moving objects other than the aether eventually stop moving.

The power of Aristotelian logic is that his ideas match our everyday experiences. Balls rolling down hills eventually stop rolling. Leaves blowing from trees fall to the ground. Water flows in a stream but stops in a pond or sea. Objects naturally fall down, as a rock from a cliff, or rise up, as smoke from fire, but do not naturally move sideways. According to Aristotelian logic, all substances undergo these transformations in motion in order to come closer to achieving their purpose. The purpose of a rock, for example, would be to seek out the place where materials made of the substance earth belong. Our experience informs us that all heavy things, given a chance, move downward. Even if we set rocks into the wall of a building, eventually that wall will collapse and the rocks will fall downward. And what does "downward" mean? To Aristotle and Aristotelians, the meaning of "downward" is obvious: toward the center of the Earth, which is the center of the universe.

What about water and air and fire? If we place water in a jar on a rocky plateau, the jar eventually will break. The water will flow downward, ponding above the rock but below the air. Air trapped under the water will escape upward. Fire from a flame will rise above the air. Each element ultimately will seek and find its natural location in the physical universe. If a material had reached its place in the universe, it would have achieved its telos and would have no further need to move. Thus, since materials made from earth, air, fire, and water all experience changes in position, they cannot yet have achieved their telos; they are not yet perfect.

Actually, as Aristotle realized, not all moving objects eventually stop moving; the stars and planets (including the Sun and Moon) are obvious exceptions to this rule. Round and round they go, ceaselessly, endlessly. Never in human memory, Aristotle asserted, had the stars or planets stopped moving. Therefore, celestial bodies, being changeless in their motions, and the stars being changeless in their brightnesses, illustrate what happens when objects have achieved perfection and fulfilled their telos. Certainly, if the stars and planets are perfect, they could not be made of any of the four imperfect, terrestrial substances; rather, a fifth element, the aether, meaning "to run always," must exist; furthermore, aether must be a material that has fulfilled its purpose. The aether is perfect and can never change; it is ungenerated, indestructible, ageless, and unalterable. Aristotle carried his argument even further: perfect objects must move perfectly. Since, according to ancient Greek mathematical logic, the circle is the most perfect two-dimensional shape, stars and planets must move along perfectly circular paths. Furthermore, since objects made of aether are perfect and unalterable, no change of any kind can occur, including the speeds of stars along these circular paths, for if a star were to accelerate or decelerate, a change in the aether would have occurred.

In this unification of Greek geometry, physics, astronomy, and philosophy, the planets became much more than mere wandering stars. They were by definition objects made of aether, objects located in the perfect, changeless celestial realm, each of which moved in a perfect circle as part of a set of nested celestial spheres, at the center of which is located the Earth. As such, the planets, the brightest of the celestial objects, illustrate for all Aristotle's followers to see that Aristotle's concept of telos was correct. With Aristotle, we now have physics wedded to astronomy, and through this unification we have learned a tremendous amount about the planets. In many ways, this is where our story begins. *We have a new idea: knowledge about the physical and philosophical principles that govern and substances that comprise the physical world here on Earth can teach us about the principles and substances that comprise the celestial realm, including the planets.* Clearly, if this idea is right, then further thinking about both the terrestrial and celestial realms might take us further along the road of knowledge.

Seven objects—Mercury, Venus, the Moon, the Sun, Mars, Jupiter, and Saturn—exist in the sky that are bright enough to detect with no measuring device save the naked eye *and* move around the sky, as measured on a mind's eye map of the stars. Circa 300 B.C.E., these are the seven planets. In the centuries following Aristotle, two major developments would affirm the necessity and importance of these seven, and only these seven, planets. The first of these was the change among Greek thinkers from descriptive to mathematical astronomy; the second was the birth of Christianity.

FROM DESCRIPTIVE TO MATHEMATICAL ASTRONOMY

Alexander the Great assumed the throne of Macedon in 336 B.C.E. and extended his father's empire eastward as far as India, conquering Persia, Egypt, Mesopotamia, Palestine, and Syria. With the Greek Empire extending to the edge of India, the center of Greek civilization moved eastward and entered a new era. By 300 B.C.E., the school and library at Alexandria, Egypt, had been created, patterned after Aristotle's Lyceum in Athens. One of the first great products of Greek thought in Alexandria was *The Elements*, written by Euclid c. 300 B.C.E., which even two thousand years later remains the basis for planar geometry. The importance of *The Elements* and of geometry in shaping Greek thought from this time forward cannot be overstated.

Aristotle's view of the universe, his cosmology, was geometrically based; however, Aristotle's cosmology was based on *concepts* of perfect circles, not on the *equations* of circles and spheres or on the

they have been in Divers Ages fucceffively afferted) being thus briefly enumerated, we fhall next give fomewhat a more ample Defcription and Illuftration of the feveral Hypothefes, and exhibit withall to the Reader's view the Schemes of each particular Syftem; beginning with the moft Ancient, that of *Pythagoras*, which is this.

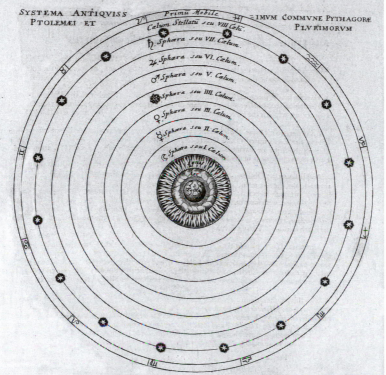

SYSTEMA ANTIQVISS
PTOLEMÆI ET

Primu Mobile
Cælum Stellatū seu VIII Cæli.
♄. Sphæra seu VII. Cælum.
♃ Sphæra seu VI. Cælum.
♂ Sphæra seu V. Cælum.
☉ Sphæra seu IIII. Cælum.
♀ Sphæra seu III. Cælum.
☿ Sphæra seu II. Cælum.
☽ Sphæra seu I Cælum

=IMVM COMMVNE PYTHAGORÆ
PLVRIMORVM

In this Syftem the **Terraqueous** Globe is feated in the midft or Centre; about it, the **Elementary** Region; next above that, the *Moon*; then *Mercury*; next above him, *Venus*; the Sun, as Moderator of all, being placed, as in a **Throne** in the midft of the Planets, environed not only by the three foregoing, called the Inferiour, but by *Mars* likewife, *Jupiter*, and *Saturn*, called the Superiour Planets. Above Saturn is the Sphere of the Fixed Stars, called 'Απλάνη, i. e. *Aplane*, or Unerring; by fome, the Firmament. The reafon of this Syftem is thus explained by *Pliny*, *Lib. 2. c. 22. Pythagoras ex Mufica ratione appellat Tonum quantum abfit a Terra Luna*, &c. i. e. Pythagoras *from Mufical Reafon*, calls the fpace between the Earth and the Moon, a Tone; the fpace from the Moon to Mercury, he will have to be half a Tone; as much in a manner from him, to Venus; from Venus to the Sun, as much and half again; but from the Sun to Mars, as much as from the Earth to the Moon, that is a Tone; from him to Jupiter, half a Tone; from Jupiter to Saturn, another half Tone; and from thence to the fixed Stars, as much and a half again. Thus are compofed Seven Tones, which Harmony they call Diapafon, that is, the generality or whole ftate of Confent or Concord. Now a Tone is by fome interpreted the fame as an Unite or Integer, and contains in Meafure according to the Opinion of *Pythagoras* 125000 *Stadia*; according to which Meafure; not only the foregoing fpace, but extent of the whole Syftem, may be computed. And this was the firft *Pythagorean* Syftem, embraced by *Archimedes*, the *Chaldeans*, *Ariftotle*, *Cicero*, *Livy*, *Ptolemy*, *Alphonfus*, *Turbachius*, and the greateft part of Aftronomers, untilthe time of *Maginus* and *Clavius*.

The

Figure 2.3: The geocentric cosmological model of Aristotle from *The Sphere of Marcus Manilius*, by Edward Sherburne, 1675. (Image used by permission of the Huntington Library)

application of mathematical geometry to the actual measured positions and motions of the planets. With *The Elements* in hand, Aristotle's descriptive astronomy rapidly gave way to mathematical astronomy. Over the next four hundred years, quantitative, mathematical astronomy would become the glue that would enable scholars to cement the grip of the geocentric universe of Aristotle onto Western thought.

Aristarchus

Aristarchus (310–230 B.C.E.), a geometer and contemporary of Euclid, was the first to attempt to measure the physical size of the universe. He measured the length of the sidereal year to be $\frac{1}{1623}$ of a day longer than 365¼ days, i.e., 365 days, 6 hours, and 53 seconds, which is just a tad shorter than the actual length of 365 days, 6 hours, 9 minutes, and 10 seconds. By recognizing that the moon reflects sunlight (that is, the moon does not shine on its own), Aristarchus realized that by measuring the angles between the Earth, the Moon, and the Sun at the moment when the Moon was exactly half illuminated, he could, in principle, determine the ratio of the distances to the Sun and Moon using basic geometric principles. Unfortunately, this measurement is exceedingly difficult to make, so his results were in error: he determined that the Sun was between 18 and 20 times farther from the Earth than the Moon (the correct answer is about 400 times more distant) and that the physical diameter of the Sun was between $\frac{38}{6}$ and $\frac{43}{6}$ times bigger than that of the Earth (the correct answer is 109 times bigger). Having deduced that the Sun was significantly bigger than the Earth, and thinking that a little object ought to go around the big object, not vice versa, he came to believe that the Earth probably revolved around the Sun. He had no supporters for this idea, though, for almost two thousand years.

In suggesting the possibility of a heliocentric rather than a geocentric universe, Aristarchus had to make the universe enormous, with the stars many times farther away than the most distant planet, Saturn. He did this in order to solve a problem first recognized by

Aristotle: if the Earth moves around the Sun within a relatively small celestial sphere in which all stars are equally distant from the center (that is, from the Sun), then the distance from the Earth to the stars on any given part of the celestial sphere will change during the year. Because the distance from the Earth to a given part of the celestial sphere changes as the Earth orbits the Sun, the observed angle between any two stars should appear to increase and decrease and increase again during the year, as our viewing position and distance to those stars change. This would have the effect of making stars appear to move back and forth, over and over again, in a cycle matched to the seasons. The only way to avoid observing this phenomenon in a Sun-centered universe is to make the size of the celestial sphere so enormous that the change in distance to the stars during the annual cycle is so tiny in comparison to the actual distance to the stars that the changes in apparent positions of the stars are too small to be measurable.

The apparent change in the position of an object because of the motion of the observer, not of the motion of the object observed, is called *parallax*, and the absence of observable parallax for the stars, with a six-month cycle, was a problem in need of an explanation. Aristotle recognized that two solutions were possible to the parallax problem: either the Earth is the center of the universe and therefore does not move, or the Sun is the center of the universe and the stars are so incredibly far away that their parallactic motion is too small to detect. For Aristotle, who lived before Aristarchus, a geocentric universe was understandable based on his physics and the idea of telos; a heliocentric universe, however, rendered his physics wrong. Why do heavy objects fall toward the center of the Earth if not because they are pursuing their telos? Why do stars move in circles if they have not achieved their telos? Therefore, in order to provide a fully consistent and understandable theory of everything, Aristotle concluded that the absence of parallax is a consequence of the Earth being at rest at the center of the universe.

Aristarchus, by virtue of putting the Earth in motion, was forced by some of the simplest laws of geometry to place the stars quite far away in order to reduce the parallax of stars to such small motions

Figure 2.4: As the Earth moves from one side of the Sun to the other, over the course of six months, the positions of some stars may appear to move relative to other stars because the position from which we view them changes. In this illustration, the Earth is shown first on one side of the Sun and then on the other. The lines from the Earth illustrate the direction toward which we look and see a particular star at those two different times of year. (Artwork by Adrienne Outlaw ©)

as to be undetectable to the unaided human eye. Aristarchus did manage to hedge his bets, though: he also constructed a geometric model for a mostly geocentric universe in which the Moon orbited the Earth in a circle in about 27 days; the Sun orbited the Earth in a circle in about 365 days; Mercury and Venus orbited in a circle around an otherwise empty point located between the Earth and Sun and along the line connecting the Earth and Sun, with that empty point orbiting in a circle around the Earth in one year; and

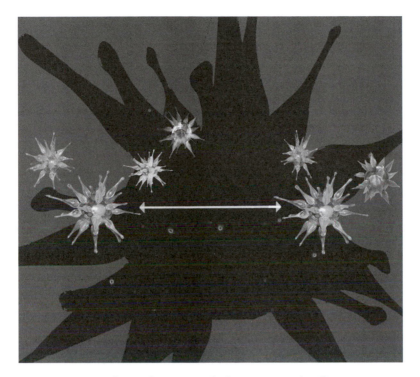

Figure 2.5: Due to the Earth's motion, the largest star in this illustration appears to move back and forth against the background of the other stars. (Artwork by Adrienne Outlaw ©)

Mars, Jupiter, and Saturn orbited the Earth in such large circular orbits that their orbital centers were near but not demonstrably identical with the position of the Earth.

Eratosthenes

The next great Greek geometer and astronomer was Eratosthenes (276–194 B.C.E.), whose greatest accomplishment was stunning: he used shadows to measure the diameter of the Earth. By making the assumption that the Sun was so far away that rays of sunlight all traveled along parallel lines, and by measuring the lengths of shadows at midday on midsummer's day at two different locations in Egypt, he deduced that the circumference of the spherical Earth was

21

250,000 stadia, which is correct to within a few percent, depending on exactly how long the measurement unit for a stadium was.

Eratosthenes also measured the distance to the Sun as 804 million stadia (correct to within 15 percent), the distance to the Moon as 780,000 stadia (about one-third of the modern value), the tilt of the direction of Earth's rotation axis to the plane of Earth's orbit around the Sun (or the Sun around the Earth) as $^{11}/_{83}$ of 180 degrees (less than one-half degree from the modern value of 23.45 degrees), and he proposed a new calendar that included a leap year every fourth year.

Hipparcus

The last and greatest of the pre–Christian era Greek mathematical astronomers was Hipparcus (190–120 B.C.E.), who lived on the isle of Rhodes and compiled a catalog of the positions and brightnesses of the 850 brightest stars in the sky, visible from the Mediterranean. Hipparcus discovered that the direction in which the Earth's rotation axis points (the north celestial pole) changes slowly; this means that today's North Star, Polaris, will no longer be the North Star in a few thousand years. What Hipparcus found was that the end point of the Earth's rotation axis, when extended as an imaginary pole outward from the surface of the Earth until it touched the celestial sphere, appears to move in a circle among the stars and that this so-called precession of the rotation axis has a period of almost 28,000 years (the correct period is close to 25,800 years). Hipparcus's measurement of this rate of precession (46 seconds of arc per year) would not be surpassed for 1,700 years (the modern value is 50.26 seconds of arc per year). Precession is the reason that the year measured based on the motion of the stars (the sidereal year) and the year based on the seasons (the tropical year) are different.

Hipparcus accurately measured the distance to the moon to be between 29.5 and 33.5 times the diameter of the Earth (the correct answer is 30) and less accurately measured the distance to the Sun to be 1,245 Earth diameters (the correct answer is 11,740 Earth diameters). Since the actual size of the Earth was known from the work of Aristarchus, Hipparcus now knew the exact physical distances to the Moon and Sun, albeit to different levels of accuracy. Hipparcus surpassed Aristarchus in accurately measuring the length of the year. He determined that the sidereal year was $365\frac{1}{4}$ + $\frac{1}{144}$ of a day long, that is, 365 days, 6 hours, and 10 minutes. He was off by only 50 seconds. He also measured the tropical year to be $\frac{1}{300}$ of a day less than $365\frac{1}{4}$ days, or 365 days, 5 hours, 55 minutes, and 12 seconds, which is 6 minutes and 26 seconds too long. Since he only claimed to have made this determination to an accuracy of 15 minutes, and since his results would not be surpassed until the seventeenth century, these results are truly astounding.

All of these accomplishments by Hipparcus and his predecessors led Hipparcus to a remarkable synthesis, a mathematically regulated geometric model for Aristotle's geocentric universe. For Hipparcus, the planets—including, of course, the Sun and Moon—were made of aether and therefore were changeless. Consequently, they had to move in perfect circles. However, unlike Aristotle, who didn't fuss with the mathematical details of his cosmological model, Hipparcus insisted that the mathematics for perfect circular motion had to lead to correct predictions (into both the future and past) for the positions of these objects at definite moments in time. Thus, from a good mathematical model, a geometer should be able to predict at a date well into the future the exact moment at which a solar eclipse or a transit of Venus should occur. Prior to Hipparcus, no astronomers had been able to develop a geometric, geocentric model that was astronomically useful for predicting future positions of celestial bodies.

In order for Hipparcus to succeed, he needed to make a few tiny adjustments to the purity of Aristotle's system. In Hipparcus's model, the Sun moved in a perfect circle; however, the center of this circle was not at the center of the Earth. Instead, it was one-twenty-fourth of the distance from the center of the Earth to the center of the Sun. He called this off-center circle an *eccentric* circle. With this geometric representation for the solar circle, Hipparcus could accurately predict future positions of the Sun.

Figure 2.6: Hipparcus's model for the orbit of the Sun. The Sun moves in a circular orbit; however, the center of the Sun's orbit was not at the center of the
Earth, as is formally required by Aristotle. Hipparcus called this off-center circular orbit an *eccentric*. (Artwork by Adrienne Outlaw ©)

The motion of the Moon is a tougher problem than the Sun.
Hipparcus solved the lunar motion problem by centering the circular
path of the Moon on a point between the Earth and Moon and
having this point revolve around the center of the Earth in nine
years. In addition, Hipparcus inclined the plane of the Moon's orbit
by five degrees to the plane of the Sun's orbit and had the entire
plane of the Moon's orbit precess with a period of nineteen years.
With this model, he could predict lunar and solar eclipses to within
an accuracy of one hour. As for the other five planets, their motions
were too complicated for Hipparcus to model.

We need to recognize that Hipparcus's geometric models for the motion of the Moon and Sun are tremendously successful mathematically; however, while they are composed of circles and circular motion, which would make Aristotle happy, they are not composed of circles centered on the center of the Earth, as should be the case in a genuine geocentric cosmology according to Aristotle. Both the lunar and solar circles are off-center from the center of the Earth, and the center of the lunar circle itself moves around a geocentric circle. So, is the Hipparcus model truly geocentric? Is it true to the philosophical demands of Aristotelian cosmology?

To make matters worse, Hipparcus found that there was a second way, mathematically, to explain the solar and lunar motions. He could obtain the same results if he put the Sun in double motion. The real Sun would be located on the circumference of a small circle, called an *epicycle*, whose center is located on the circumference of a larger circle. This larger circle was exactly and perfectly centered on the center of the Earth. Over the course of a year, the Sun moved around the circumference of the epicycle exactly once while the center of the epicycle moved along the circumference of the large circle exactly once. Similarly, the motion of the Moon could be represented by epicyclic motion, but with the center of the large circle off-center from the Earth and moving around the Earth in nine years.

If there is more than one way to represent the lunar and solar motions mathematically, which is right? Or is neither? If the mathematical astronomy works mathematically but does not strictly match Aristotle's philosophical model for the structure of the universe, which model does one accept as right? Are our mathematics simply imperfect human constructs? Are we intellectually incapable of constructing the "true" mathematics that would reveal and explain the real motions of the perfect Aristotelian universe? Hipparcus preferred his epicyclic model for lunar motion because he believed that it was more strictly Earth-centered, more true to Aristotelian cosmology. But is it? These questions and problems would haunt astronomers for the next 1,700 years.

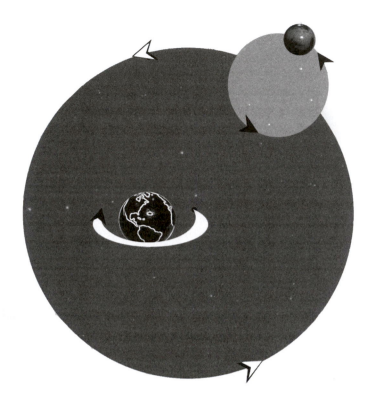

Figure 2.7: Hipparcus's second model for the orbit of the Moon. In this model, the Moon moves on a small circle—an epicycle—whose center is on an eccentric, a large circle that is not centered on the Earth. The Moon orbits the center of the epicycle while the center of the epicycle orbits the center of the eccentric and the center of the eccentric orbits the Earth. (Artwork by Adrienne Outlaw ©)

Ptolemy

Klaudios Ptolemaios (c. 85–165 c.e.), known to us as Ptolemy, was without question the most important and influential of the Greek astronomers, as his synthesis of Aristotelian physics with technical, mathematical astronomy produced a worldview known as the Ptolemaic universe, which is a complicated, mathematicized version of Aristotle's geocentric universe. For one and a half thousand years, from the second through the sixteenth centuries, Ptolemy's astronomy dominated scientific thought, and his most important book, *Almagest*, defined astronomy and astronomy education in both Christendom and the Islamic world. Ptolemy's greatest accomplishment was finding a way to mathematically model the motions of all seven planets, not just the motions of the Sun and Moon. For the Sun and Moon, his model was almost indistinguishable from that of Hipparcus, except that for both objects Ptolemy preferred single, off-center circles rather than epicycles. He did so because he viewed a one-circle model as simpler than a two-circle model in that it requires one motion, not two. We should note, however, that the one-circle model is not simpler mathematically, and thus simplicity, like beauty, is in the eye of the beholder.

The problem with the orbits of the planets, not including the Sun and Moon, is that their motions are highly irregular when viewed over long periods of time. All the planets share one motion, the east to west diurnal rotation of the sky, with the stars. In addition, like the Sun, the planets all have a second motion, a regular movement from west to east, in the opposite direction as the general motion of the stars. We call this motion of the planets from west to east *prograde motion*. But every time either Mars, Jupiter, or Saturn approaches opposition, which occurs when an object is seen in nearly the opposite direction in the sky as the Sun, its motion through the background stars appears to slow down and eventually stop going prograde; the planet then reverses course, speeds up and moves *retrograde* (i.e., in the same direction as the stars), slows down and stops, and reverses course again. Mercury and Venus do the same thing; however, Mercury and Venus are never seen at opposition, since they always stay near the Sun. Hence, they experience retrograde motion when they are close to inferior conjunction, that is, nearly in the same direction as the Sun and in between the Earth and Sun. One cannot help but notice that all five planets that show retrograde motion (the Sun and Moon do not) do so in a way that appears to be associated with the Sun; specifically, retrograde motion occurs close to the line connecting the Earth and Sun, as this line maps out points of conjunction and opposition.

If the speed of a planet moving on an epicycle is adjusted correctly, epicyclic motion can be used to produce retrograde motion. Hipparcus knew this and came close to producing mathematical models of the planets' motions using carefully calibrated epicyclic speeds; however, these simple epicyclic models did not quantitatively predict the positions of the planets, as demanded by Hipparcus, and so he abandoned his efforts to model planetary motion. Hipparcus's legacy is not just that he created successful mathematical models for the motions of the Moon and Sun, but that he demanded that future astronomers meet his standards of accuracy for models of the orbits of the other planets. In this way, he defined the science of astronomy.

In order to meet the rigorous standards of Hipparcus, Ptolemy invented another tool for the mathematical astronomer's tool kit. He invented the *equant*. The equant is a strange but brilliant concept, yet it is nothing more than a point in space around which a planet's orbital speed is measured. An object's orbital speed can be measured in two different ways. The first way is our normal concept of physical speed: distance traveled in a defined unit of time. When we travel by car, for instance we might travel at 100 km per hour. In ten hours we would travel 1,000 km. The second way to measure speed for an object moving in an orbit is *angular velocity*. Given that a circle encompasses 360 degrees, an object moving 36 degrees per hour would go around a circle once in ten hours. Note that for angular velocity, we have no idea what the physical speed of the moving object is unless we know the actual size of the orbital circle. If the circle is large, the object moving at 36 degrees per hour would have a much greater physical speed than if the circle were small. For example, if the circumference of the circle is 30,000 km (nearly the circumference of the Earth at the equator), the physical speed of an object moving 36 degrees per hour would be 3,000 km per hour; on the other hand, if the circumference were only 10 km, the physical speed of an object moving 36 degrees per hour would only be 1 km per hour.

For an object in a circular orbit, constant physical speed and constant angular velocity produce the same result: the orbiting body

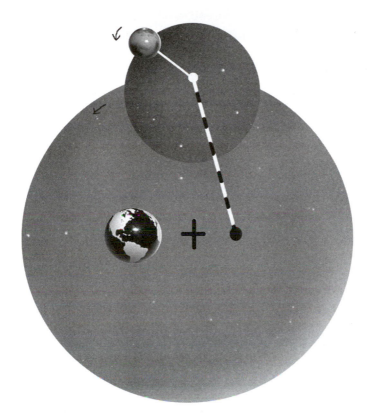

Figure 2.8: Ptolemy introduced the equant point. This point is as far as the Earth from the center of the eccentric circle of an orbit, but on the opposite side of the center of the eccentric from the Earth. The center of the planetary eccentric moves at a constant angular speed, as seen by an observer positioned at the equant point. (Artwork by Adrienne Outlaw ©)

moves on a circular path with neither the physical speed nor the angular velocity ever changing. The planets, however, were well known to the ancients to have changing angular velocities. After all, they not only change angular velocity by slowing down and speeding up, they actually change directions when they switch from prograde to retrograde motion.

The planets also were well known to change in apparent size and brightness. Typically, the planets appear biggest and brightest

near opposition (Mars, Jupiter, and Saturn) or near inferior conjunction (Mercury, Venus). One obvious way to affect the size and
brightness of a planet is to adjust the distance between the planet
and the Earth. For example, if the distance to an object doubles,
both the apparent size and the brightness of the object decrease by
a factor of four. One can change the actual distance to a planet by
putting it on a circular orbit for which the center is not the center
of the Earth. At one side of the orbit, the planet is closest to Earth,
making it appear bigger and brighter than average to a viewer on
the Earth; at the other side of the orbit, the planet is farthest from
Earth, making it appear smaller and fainter than average. Hipparcus
and Ptolemy both put the Moon on a circular orbit, off-center from
the Earth. With the Moon moving at constant physical speed and
constant angular velocity, they could reproduce the motion of the
Moon to a high degree of accuracy and the changing size and brightness of the Moon to a lesser degree of accuracy. But this method will
not produce retrograde motion; hence, Ptolemy was forced to invent
something new.

We need to imagine a circular orbit with the Earth off-center.
Now, we draw a line from the Earth to the center of the circle and
measure that distance. Next, we continue our line from the Earth
to and through the center, ending our line at a distance equally far
on the opposite side of the center as the Earth is on its side. This
point, equally distant from the center as the Earth, but on the opposite side, is the equant.

Ptolemy's trick was to require that a planet move at constant
angular velocity as seen from the position of the equant. In contrast,
as seen from the center of the circle or from the position of the
Earth, the angular velocity of the planet is not constant. If the planet
is on an epicycle, Ptolemy required that the angular velocity of the
planet around the center of the epicycle, as seen from the center of
the epicycle, was constant while the angular velocity of the center
of the epicycle was also constant, as seen from the position of the
equant. The beauty of this absolutely bizarre innovation is easy to
appreciate: it works. A planet on this kind of orbit is biggest and
brightest and moves with the greatest angular velocity—as seen from

the Earth—and also experiences retrograde motion when it is closest to the Earth. With the right set of parameters (sizes of the orbit and the epicycle, speed of motion on the epicycle and by the center of the epicycle), Ptolemy reproduced mathematically the orbits of all the planets. His model of the heavens also included measurements of the distances to all seven celestial wanderers.

Ptolemy viewed his work as that of an Aristotelian: he believed he had preserved Aristotle's concept of perfect circular motion. He also had loosely preserved Aristotle's concept of changelessness of the heavenly bodies, due to their perfection, through his use of constant angular velocity measured from the equant. In addition, the heavenly bodies did not change in size or brightness, they only appeared to do so as viewed from the Earth. Ptolemy created a synthesis of technical, mathematical astronomy and Aristotle's geocentric universe. We, however, might think of Ptolemy as a very cavalier Aristotelian. After all, none of the planets' orbits actually are centered on the Earth; each planet has its own equant point, and as seen from the Earth the orbital speeds of the planets are constantly changing, rather than being constant. But by preserving the concept of a geocentric universe built upon circles while building a mathematically successful model, Ptolemy had saved the appearances of Aristotelian cosmology. Thus, the Ptolemaic universe took root and would enable Aristotle's metaphysics, including his idea of a geocentric universe, to dominate scientific thought for another one and a half millennia.

THE IMPACT OF THE BIRTH OF CHRISTIANITY ON ASTRONOMY

Many early Christian theological debates involved showing how certain Christian concepts could be seen as consistent with critical aspects of the natural philosophies of Aristotle and Plato. Aristotle's concept of the perfect heavens centered on the Earth was seen as not only compatible with but also inseparable from Christian beliefs. The perfect heavens were the abode of God and the angels, with angels needing wings to fly up to the heavens; the central Earth was imperfect, unlike the perfect aether of the heavens, and therefore the Earth was the natural locus of evil and sin. By late medieval times, the Aristotelian synthesis begun in the thirteenth century by Albertus Magnus and capped by the work of his brilliant student Thomas Aquinas secured an importance in medieval scholarship for the ideas and writings of Aristotle second only to that of the Bible.

Aristotle developed a cosmology complete with seven celestial bodies, outside of which was the celestial sphere itself. While Aristotle explained the presence of seven planets, in later times Christian theologians, enamored by the number seven, would argue that seven and only seven planets could exist, no more and no less. In his play about Galileo Galilei, *Lamp at Midnight*, playwright Barrie Stavis illustrates this issue in a scene in which Galileo attempts to have three close friends in Florence peer through his newly invented telescope to see the four moons Galileo has discovered around Jupiter:

Professor Magini, Chief Mathematician at the University of Bologna.
So, Galileo, you insist that four moons revolve around the planet Jupiter?
Galileo. I do.
Magini. That means you want to increase the number of planets from seven to eleven?
Galileo. I have nothing to do with diminishing or increasing their number. I've fastened my telescope on Jupiter night after night and I have discovered that just as our earth has one moon revolving around her, so Jupiter has four moons. *This* night I want you to look through that telescope and see them for yourself.
Professor Sizzi, a monk and Professor of Astronomy from Florence. But Aristotle teaches that there are seven planets.
Galileo. Aristotle died two thousand years ago. He knew much—but not everything. He might have been wrong.
Professor Libri, Professor of Philosophy at the University of Pisa (horrified): Aristotle wrong?
Magini. Are you denying Aristotle's wisdom?

Galileo. To say that two thousand years ago one man knew everything there was to know is to deny the future. How can we set bounds to man's understanding? Who dares assume that everything that can be known in the world has already been found out?

Sizzi. There are only seven planets in the sky! Aristotle said so. I'll show you the page. (He takes a large book from Libri, who has brought it with him, and turns the pages to find the place.)

Magini (looking over Sizzi's shoulders). There!

Libri (looking over Sizzi's and Magini's shoulders). There! (Triumphantly, they hold the book open before Galileo.)

Galileo (shutting the book without looking at it). And what will you say when you look through this telescope and see eleven planets—with your own eyes?

Magini. I would not believe my eyes. Their existence is contrary to the principles of common sense.

Sizzi. I don't need to look. Seven. Seven! No more—no less! Aristotle said it.

Galileo. If all science were contained in Aristotle, you three gentlemen would be the greatest scientists in the world—you have every passage at your fingertips.

Magini (patiently). Galileo, the world is founded on the number seven.

Libri. Seven is a *holy* number.

Galileo. For the mathematician every number is holy—none more so than any other.

Magini. Both nature and Scripture show that seven is holy. The seven metals of the earth. The seven golden candlesticks of the Apocalypse. The seven intervals in the octave of music.

Sizzi. The seven churches of Asia. The seven—

Galileo (ironically). The seven mortal sins.

Sizzi (looks at him sourly). The seven Penitential Psalms.

Magini. And the seven days of the week named for the seven planets. Increase the number of planets and the whole system falls to the ground. Are we to have eleven days to the week? Eleven days to the week!

Sizzi. How many windows are there in the head? Seven! Two nostrils, two eyes, two ears, one mouth. So in the heavens, two favorable stars, two unfavorable, two luminaries, and Mercury, undecided and indifferent.

Magini (sagely). Seven is undoubtedly a holy number.[1]

Medieval scholastics found further evidence that the number seven is holy in the seven days of creation, the scriptural requirement of a seven-year cycle for planting crops (in which fields were to lie fallow in the seventh year), the seven years of plenty followed by seven years of famine found in the story of Joseph in Egypt, the seven years needed for Solomon to build the ancient temple in Jerusalem, and the supposed seven gates of heaven. Further, obvious support for the importance of the number seven came from the recognition that there were seven wonders of the ancient world (the Great Pyramid of Giza, the Hanging Gardens of Babylon, the statue of Zeus at Olympia, the Temple of Artemis at Ephesus, the Mausoleum at Halicarnassus, the Colossus of Rhodes, and the Lighthouse of Alexandria), the seven seas (North Atlantic, South Atlantic, Indian, North Pacific, South Pacific, Arctic, and Antarctic oceans), and the seven hills of ancient Rome (Palatine, Capitoline, Quirinal, Viminal, Esquiline, Caelin, and Aventine).

Aquinas explained that there were not only seven celestial bodies but also seven heavens. The innermost four heavens, for Aquinas, were in the imperfect, sublunary realm. The lowermost of these was the airy heaven, or lower region of air, above which was the ethereal heaven, or the upper region of the sphere of air. Above these two came two heavens within the sphere of fire, first the Olympian heaven found at the approximate height of Mount Olympus, and next the fiery heaven, or the upper region of fire. The fifth heaven, being the lowermost heaven in the perfect celestial realm, was the sidereal heaven, which contained all seven planetary spheres and the celestial sphere. Nested just beyond the sidereal heaven was the crystalline, or aqueous, heaven. This heavenly sphere, which was starless and transparent, was made of crystallized water, similar to pure and completely transparent ice. The crystalline heaven was believed required by the lines in Genesis 1, in which the waters above were separated from the waters below, such that there were waters above the dome of the sky. Finally, the outermost heaven, the seventh heaven, was the Empyrean heaven. The seventh heaven, filled with light and angels, was brilliant in its fiery and intellectual splendor. It provided permanence and stability on all the spheres below and was the abode of God. One can easily imagine why one would like to rise, or be taken, upward to the seventh heaven.

To medieval scholars, the order by distance of the seven planets, as arranged by Ptolemy—the Moon, Mercury, Venus, the Sun, Mars, Jupiter, and Saturn—also made clear and natural sense. The Sun, unique as it is, divided the three similarly behaved outer planets from the differently behaved inner planets. Thus, wrote Themon Judaeus, the Sun was "the wise king in the middle of his kingdom, and as the heart in the middle of the body."[2] To add more planets would render this seven-planet hierarchy wrong and befuddle over a thousand years of intellectual effort by Christian scholars devoted to understanding God's word in the context of Aristotelian natural philosophy.

Thus, as intellectual activity in the Western world begins to accelerate in the late Middle Ages, we have seven planets. Not only are seven planets found in the sky, scholars of the era also provided powerful philosophical, metaphysical, and theological reasons why there should be and *must* be seven planets. Such is the word of Aristotle and, as once understood by Christian interpreters, such is the Word of God.

3
The Earth Becomes a Planet

O

In the late Middle Ages, leading scholars began to challenge the intellectual dominance of Aristotle; Aristotle's hegemony in physics and astronomy, however, was nearly the last to succumb to the forces of change. The physical cosmology of Aristotle withstood the challenges well, in part because it was built on Aristotle's apparently irreplaceable laws of physics and the universally accepted concept of telos and in part because the centrality of the Earth in the universe was an important pillar of orthodox Christian theology. Only after three centuries of increasingly successful challenges to Aristotle in other areas of thought were astronomers able to displace the Ptolemaic view of the universe with a heliocentric model of the cosmos. With this change in our cosmological view, the Earth switched places with the Sun, with the Earth becoming a planet and the Sun assuming the exalted position at the center of the known universe.

THE PARIS TERMINISTS

In the fourteenth century, Jean Buridan (c. 1320 c.e.), Nicole Oresme (c. 1325–1382 c.e.), and the Paris Terminists challenged Aristotle's ideas on motion. According to their "impetus" theory, a body set in motion receives an impetus from the body or thing that pushes it. After receiving this impetus, the moving body would continue to move forever, without need of another push, were there no air resistance to slow it down and bring it to a stop. If one applies this concept to heavenly bodies, Aristotle's concept of telos, indeed the existence of a fifth essence, the aether, would be unnecessary. As Buridan wrote, "[I]t could be said that when God created the celestial spheres, He began to move each of them as He wished, and they are still moved by the impetus which He gave to them because, there being no resistance, the impetus is neither corrupted nor diminished."[1]

Oresme based a further argument on the philosophical principle we recall as Ockham's razor: what can be done with fewer assumptions is done in vain with more. William of Ockham (1280–1349 C.E.) was a critic of knowledge. Buridan studied under Ockham, Oresme under Buridan. The upshot of Ockham's logic is that simple explanations are preferable to complicated ones: a theory should be no more complicated than absolutely necessary. Furthermore, if two explanations are equally good at explaining an observed phenomenon and neither one is demonstrably more correct than the other, the simplest explanation is best.

Oresme's powerful new view of the cosmos, if accepted, would reverse the motions of the heavens. He suggested that rather than have the Earth sit still while the celestial spheres wheel around the Earth, the celestial spheres might be fixed while the Earth rotates once a day. Both motions would produce the same appearance of the stars rising and setting. "It seems to me," wrote Oresme, "that we might well affirm, subject to correction, . . . that it is the Earth that makes a daily rotation, and not the heavens. And I would like to assert the impossibility of establishing the contrary claim first by means of any observations, or secondly, by means of any rational process."[2] We should note, though, that a spinning Earth can explain the daily motions of the stars; however, Oresme's model also requires the Earth to revolve about the Sun in order to account for the yearly changes seen in the heavens.

Can one argue that one view is simpler than the other and use Ockham's razor to select that as the best answer? A geocentric universe demands two motions: a spinning celestial sphere to explain day and night and a moving Sun to explain the cycle of the seasons. A heliocentric universe also demands a double motion: an Earth spinning on its axis for day and night and an Earth revolving about the Sun for the annual cycle. Which is simpler? Oresme argued that both views are equal in their relative simplicity, or lack thereof, and thus rational arguments prevent one from choosing between them; therefore, he chose the geocentric universe since that one was believed to be compatible with scripture, as interpreted by medieval Christian theologians.

POLITICS, MEDICINE, AND RELIGION

A century later, Aristotle was dethroned in several spheres of thought. Nicolo Machiavelli (1469–1527 c.e.) challenged the Aristotelian view that people are inherently virtuous, with *virtuous* taken here to mean morally excellent, that the goal of political leadership is to produce a just society, and that producing such a society was the telos for a political leader. Rather, to Machiavelli, a virtuous politician is a successful ruler, and he is successful if he maintains power to ensure the security, wealth, and well-being of his city-state. Virtue in a politician, to Machiavelli, had no necessary connection to moral excellence. Machiavelli assumed that all men are venal and that a good ruler, rather than trying to raise men up to become more moral, as would an Aristotelian leader, should manipulate and control the populace by exploiting their selfish natures. When Machiavelli wrote *The Prince*, which was published posthumously in 1533, he believed he was describing effective politics and successful rulers as they were, not as Aristotle believed they should be. The political philosophy of Aristotle (and Plato) was swept away in a torrent of political upheaval in the sixteenth century, supported by Machiavellian principles, leaving an intellectual landscape upon which Aristotle's hegemony was slightly diminished.

As taught in medieval universities since the University of Bologna was founded in 1156 c.e., the art and practice of medicine continued to be based on an understanding of the human body according to Galen. Galen, a Roman physician born in Asia Minor in 129 C.E., did internal studies on animal and human bodies. He used his studies to expand on the ideas of Hippocrates (460–370 B.C.E.) in order to develop a model for how the human body functions. Just as the terrestrial part of the Aristotelian universe contained four elements, Hippocrates premised his understanding of the body on the existence and balance or imbalance of four fundamental elements: blood, phlegm, yellow bile, and black bile. To Hippocrates, diseases were due to natural imbalances between these four bodily fluids. Five hundred years after Hippocrates, Galen retained the four elements and even identified the physical locations of three souls— the animal, vegetable, and rational souls—inside the body. Aristotle's philosophy, and therefore his physics and cosmology, was based on the existence of these three souls, and since the time of Galen, Hippocratic medical practice was understood in terms of Aristotle's tripartite soul. If the rational soul, which enables us to use logic to discover the laws of the universe, can be located within an organ inside the body, then it obviously exists. Because the rational soul was critical to early Christian philosophy as well, Galen's medical views, like Aristotle's geocentric universe, calcified in a society in which intellectual thought was exclusively in the service of the Church of Rome.

Any challenge to Galen would be a challenge to both Aristotle and the established theology of the Catholic Church. Dissection in the service of medicine almost certainly would lead to an improved understanding of how the body worked and thus would lead to better understandings of illness and the practice of medicine. Though dissection became legal in Italy in 1405, the church continued to forbid dissections by clergy; and since, by religious edict, all medical students and physicians were clergy, little learning about the human body occurred in the fifteenth century. As a result, anatomy as taught and understood in 1500 was identical to that presented 1,400 years previously by Galen. That is, this was the status quo until the work of Andreas Vesalius (1514–1564 C.E.).

By the time Vesalius studied medicine, in the 1530s, all the original writings of Galen had been translated from Arabic into Latin.

To Vesalius, the obvious errors in Galen's understanding of the human body could no longer be ascribed to the absence of adequate translations of some of Galen's work. Secretly, Vesalius broke with tradition and performed dissections. Despite the authority of Aristotle, who identified the three parts of the soul with three internal organs, Vesalius could not find or identify the three souls with individual organs inside the body. In just that nondiscovery, the Aristotelian premise upon which the practice of medicine was built collapsed. As for theology, in contradiction to the biblical expectation that Eve had been fashioned from one of Adam's ribs and therefore that men had fewer ribs than women, Vesalius found that men and women have the same number of ribs, a finding that presented an obvious challenge to orthodoxy. In 1543, he published his monumental work *De Humani Corporis Fabrica* (On the Fabric of the Human Body), a five-hundred-page book with 250 block print illustrations based on what he saw, not what Galen said he should see, inside a human body. Like a virus, his book spread quickly throughout Europe. Vesalius's work was so clearly, so obviously, superior to that of Galen that within a generation it came to dominate medical education. Thus, the preponderance of Aristotelian thought within the intellectual sphere became, again, slightly less overwhelming.

While Aristotle was challenged directly through new thought and progress in politics and medicine, Martin Luther presented another kind of challenge to authority. Launched unintentionally by Luther on the thirty-first of October 1517, when he posted his Ninety-five Theses on the door of the Castle Church in Wittenberg, the Protestant Reformation rolled across northern Europe in barely a decade. Medieval Catholicism stressed a human relationship with God that was based on using reason to discover God. In this sense, medieval Catholicism was very strongly Aristotelian. Through logic and reason, medieval theologians developed ideas about a *rational, comprehensible* universe in which salvation was based on a man's life, his works, and the relationship he develops with a rational God.

Luther reversed this: man is not saved based on his works. According to Luther, man's own efforts to achieve salvation are ignored by God. Man cannot merit salvation. He is simply given (or not given) salvation by a just and loving God. Man cannot persuade or understand an omnipotent God, and therefore using reason to try to understand God is presumptuous. Reason, he argued, is a human construct and has no place in theology. Almost overnight, in this most important of medieval activities, religion, the Aristotelian basis of Christianity had been overthrown.

In contrast to earlier centuries, during which challengers like Oresme and Buridan rose up against Aristotle but failed to find sufficiently powerful arguments to displace him, the intellectual fervor of the sixteenth century in politics, medicine, and religion all led to successful challenges to Aristotelian-based thinking. Aristotle's ideas could be criticized and replaced with new ways of thinking.

With these changes in the intellectual climate in mind, we now turn back to astronomy where we can once again take up the question *What is a planet?* As we shall see, the answer to this question is about to change dramatically through the events known as the Copernican Revolution.

ASTRONOMY

Nicholas Copernicus was born in Torun, Poland, in 1473. With the support of his uncle, a Catholic bishop, Copernicus studied first at the University of Cracow. He then went to the University of Bologna to study canon law but instead fell under the influence of the mathematician and astronomer Domenica Maria de Novara. Having failed to complete his canon law studies at Bologna, he began (but did not complete) the study of medicine at the University of Padua and finally completed his legal studies at the University of Ferrara. He returned to Frauenberg in 1503 to a comfortable administrative position as a canon of the cathedral, where he remained for the last forty years of his life.

About 1510, Copernicus composed but did not publish a short essay, *De Hypothesibus Motuum Coelestium a se Constitutis Commentariolus* (A Commentary on the Theories of the Motions of Heavenly Objects from Their Arrangements), in which he pre-

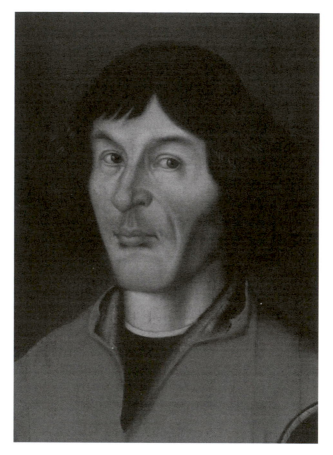

Figure 3.1: Sixteenth-century portrait of Nicholas Copernicus. (Reproduced with permission of the Regional Museum in Torun, Poland)

sented seven postulates. These postulates were the foundation of a new, heliocentric cosmology, though perhaps it would be better to say that these postulates reconstructed the heliocentric idea that had been dormant since it was first proposed by Aristarchus in the third century B.C.E. more than 1,800 years before. Two millennia is a long time for an idea to remain dormant, but such was the dominance of Aristotelian ideas.

According to Aristotle's view of the universe, the terrestrial realm of earth, air, fire, and water in which we live was one "world," with one set of physical laws, while the celestial realm of aether was a second, separate "world" in which an entirely different set of physical laws applied. According to the physical laws that govern the celestial realm, all celestial objects necessarily move in perfect, ceaseless, changeless, circular paths around the center of the universe, that center being located at the center of the Earth. Not only did Aristotelian physics declare that the only objects that can possess perfect circular motion are those objects made of aether and that those objects must be in the heavens, Aristotelian physics decreed that all objects made of aether must possess the trait of circular motion around the center of the universe.

Copernicus's principles (numbers 1 through 7, below, with explanations in brackets)[3] reversed the places of the Earth and Sun; in doing so, Copernicus put the Earth into circular motion even though the Earth was made of earth, not aether, and it stopped the Sun from moving, even though the Sun was made of aether. Although more than a century would pass before Isaac Newton would provide a unified set of physical laws that would govern both the celestial and terrestrial realms, Aristotelian physics was clearly incompatible with a Copernican universe.

1. There is not a single center for all the celestial orbs or spheres. [The Sun is one center of motion, for the planets; the Earth is a second center of motion, for the Moon.]

2. The center of the Earth is not the center of the world [the Sun is the center of the world], but only of the heavy bodies and of the lunar orb [rocks fall toward the center of the Earth, and the Moon circles the Earth].[4]

3. All the orbs encompass the Sun which is, so to speak, in the middle of them all, for the center of the world is near the Sun. [The planets have orbits or spheres centered on the Sun.]

4. The distance from the Sun to the Earth is insensible [so small as to be not measurable] in relation to the height of the firmament [the stars are much, much more distant than the Sun].

5. Every motion that seems to belong to the firmament does not arise from it, but from the Earth. . . . the firmament, or last

heaven, remains motionless. [This is Oresme's idea: the stars appear to move because of the Earth's motion, but in fact the Earth turns while the stars remain still.]

6. The motions that seem to us proper to the Sun do not arise from it, but from the Earth and our orb, with which we revolve around the Sun like any other planet. In consequence, the Earth is carried along with several motions. [The Sun is still. The Earth experiences two fundamental motions: rotation on its axis, which gives us day and night, and revolution around the Sun, which gives us the year and the seasons. A third motion, known as precession, accounts for the 26,000-year period discovered by Hipparcus. The Earth is treated "like any other planet."]

7. The retrograde and direct motions which appear in the case of the planets are not caused by them, but by the Earth. The motion of the Earth alone is sufficient to explain a wealth of apparent irregularities in the heavens. [Retrograde motion occurs when the Earth "laps" an outer planet or an inner planet "laps" the Earth. The planets do not actually slow down, stop, and reverse directions when they are seen to go retrograde.]

Like Any Other Planet

By tossing aside the geocentric universe and placing the Sun at the center of the cosmos, Copernicus has made the Earth a planet, just like any other planet. The Sun, just as surely, is no longer a planet, while the Moon has metamorphosed into an ill-defined object that is, for the time being, unique in that it orbits a planet whereas planets orbit the Sun.

Although *Commentariolus* remained formally unpublished until the nineteenth century, circulated privately by Copernicus in the years after 1510 but only among his friends and acquaintances, the ideas therein gained widespread general circulation in Europe, and by 1515 Copernicus and his ideas were famous. In fact, in 1516, the pope requested help from the renowned astronomer Copernicus in reforming the calendar, which authorities recognized was insufficiently accurate for the long-term needs of the Catholic Church.

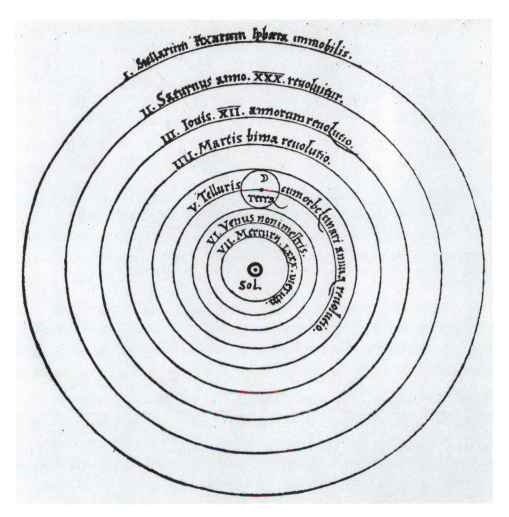

Figure 3.2: The heliocentric cosmology of Copernicus. From *De Revolutionibus Orbium Celestium*, 1543. (Image courtesy of Image Select / Art Resource, NY)

The problem with the calendrical year used in Europe was that it did not account for the 11¼ minute difference between the tropical year (365.2422 days) and the average calendar year (365¼ days). These extra 11¼ minutes, each year, rapidly accumulate into almost a 1-hour offset in only 5 years and into a full 24 hours in only 128 years. The consequence of the 11¼ minute difference was that astronomically inflexible events, such as the winter and summer solstices (the days of the year when we have the fewest and most hours of daylight, respectively) and the vernal and autumnal equinoxes (the days of the year when every part of the Earth has 12 hours of daylight and 12 hours of night), were migrating through the calendar. Thus, as determined by astronomical measurements, the first day of spring, the vernal equinox, had slipped from March 21 to March 11. As measured by the changing seasons, March 21 was moving from spring toward summer.

The date of Easter was not a fixed day on the calendar but was calculated anew for each year as the first Sunday after the first full moon after the vernal equinox; in addition, church authorities, without regard for the actual astronomical events which they could not control, arbitrarily fixed the date of the vernal equinox as March 21. In contrast, Christmas was a fixed date (December 25) on the calendar. Consequently, Easter, which always occurred within a few weeks after March 21, was migrating forward through the seasons from spring toward summer while true spring was marching backward toward Christmas. Eventually, Easter would occur in the midsummer and Christmas would occur in the spring. Meanwhile, Easter was losing its special relationship with the Jewish holiday of Passover, as Passover was moving through the Latin calendar but staying fixed on the Jewish calendar as a spring holiday. Eventually, Passover would be closer to Christmas than to Easter. Such an eventuality was unacceptable to Catholic authorities. The calendar needed to be fixed. But despite his fame and pressure from the highest levels in the Vatican, Copernicus refused to help in 1515, saying that his mathematical tools and models were not yet ready for that task. (Eventually, the calendar used in Catholic countries would be fixed, but not until 1582.)

A year later, the Protestant Reformation began, making the next few decades an inauspicious time for challenges to authority from within the parts of Europe remaining loyal to the authority of Rome. Beginning in 1536, some of Copernicus's friends and associates began urging him to publish his ideas. Finally, when literally on his deathbed in 1543, he completed this task, publishing the monumental work *De Revolutionibus Orbium Celestium* (The Revolutions of the Celestial Orbs).

One might think that the astronomy community, and perhaps all educated persons within earshot of the influence of Copernican astronomy, would have made a quick transition from the idea of a geocentric universe with seven planets to a heliocentric universe with six planets (including the Earth) and one Sun and one Moon. This did not happen, however. By the end of the sixteenth century, the total number of astronomers who agreed with Copernicus could be counted on a few fingers of one hand.

Why? The idea that the Earth moved, that the Earth was not at the center of the universe, was in contradiction both to Aristotelian physics and to the established views of the holy fathers of the Roman Catholic Church. To mid-sixteenth-century thinkers, putting the Earth in motion simply made no sense. Jean Bodin, a political philosopher who opposed Machiavelli's views, and whose writings were all placed on the Index of banned books in 1628, wrote:

> No one in his senses, or imbued with the slightest knowledge of physics, will ever think that the Earth, heavy and unwieldy from its own weight and mass, staggers up and down around its own center and that of the Sun; for at the slightest jar of the Earth, we would see cities and fortresses, towns and mountains thrown down. . . . [A]ll things on finding places suitable to their natures, remain there, as Aristotle writes. Since therefore the Earth has been allotted a place fitting its nature, it cannot be whirled around by other motion than its own.[5]

Without new laws of physics to replace those of Aristotle, without any explanation as to *why* celestial objects might orbit the Sun rather than the Earth, most astronomers, as well as most physicists, natural philosophers, and theologians, remained unconvinced and even opposed to Copernican cosmology.

The Florentine Aristotelian philosopher Ludovico delle Colombe summarized the biblical verses often cited as proof that

Copernicus was wrong in his 1611 treatise *Contro il moto della terra* (Against the Motion of the Earth). The anti-Copernican, pro-Aristotelian scriptural citations pointed out by Colombe include:

> *Psalms 104:5*: You fixed the earth on its foundation, never to be moved.
>
> *Job 104:7*: He stretches out the North over empty space, and suspends the earth over nothing at all.
>
> *1 Chronicles 16:30*: Tremble before him, all the earth; he has made the world firm, not to be moved.
>
> *Isaiah 40:12*: Who has cupped in his hand the waters of the sea, and marked off the heavens with a span? Who has held in a measure the dust of the earth, weighed the mountains in scales and the hills in a balance?
>
> *Proverbs 27:3*: Stone is heavy, and sand a burden, but a fool's provocation is heavier than both.
>
> *Ecclesiastes 1:5*: The sun rises and the sun goes down; then it presses on to the place where it rises.

A seventh piece of scripture not noted by Colombe but that is by far the most famous piece of scripture cited in opposition to Copernican cosmology is from Joshua:

> *Joshua 10:12–13*: On this day, when the Lord delivered up the Amorites to the Israelites, Joshua prayed to the Lord, and said in the presence of Israel: Stand still, O sun, at Gibeon, O moon, in the valley of Aijalon! And the sun stood still, and the moon stayed, while the nation took vengeance on its foes. Is this not recorded in the Book of Jashar? The sun halted in the middle of the sky; not for a whole day did it resume its swift course.

According to Roman Catholic Church authorities, scripture did not have to be interpreted literally; however, the job of interpreting scripture rested with the church fathers, not with individuals. More to the point, these passages had been interpreted by respected church authorities as being consistent with a geocentric cosmos.

On April 8, 1546, at the Council of Trent, which convened from 1545 until 1563 to develop strategies to counter the Lutheran reformers, the Roman Catholic Church established the doctrine that no one is allowed to "twist the sense of Holy Scripture against the meaning which has been and is being held by our Holy Mother Church whose duty it is to judge of the true sense and interpretation of the Holy Scriptures, nor shall any one dare to interpret these Scriptures contrary to the unanimous consensus of the Fathers, even if such interpretations have never before been brought to light."[6] This doctrine directly opposed the Lutheran assertion that every individual was empowered to read and interpret scripture.

The interpretations of these biblical verses dealing with the physical relationship between the Earth and Sun had been made in the post-Aquinas era during which Catholic theology had become wedded to Aristotelian philosophy and physics. Between 1544 and 1547, the Dominican friar and astronomer Giovanni Maria Tolosani composed a refutation of heliocentrism based on physical, scriptural, and philosophical arguments.[7] As viewed from Rome, these Bible verses were compatible with Aristotle's view of the universe because Aristotle had correctly understood the physical universe as created by God. The Roman Catholic Church quickly took up a position opposing any challenges to their authority, and certainly heliocentrism was such a challenge.

Copernicus also was directly challenged by the leading Protestant reformers, including Luther, Calvin, and Melanchthon, all of whom opposed heliocentrism because it contradicted several pieces of scripture, and scripture (as interpreted by them) was the only source of knowledge (so they claimed) for a believing Christian. Armed with literal translations of these verses, Protestant reformers had all the ammunition they needed to attack Copernicus and his ideas. In 1539, before Copernicus had even published *Orbium Celestium*, Luther reportedly said:

People gave ear to an upstart astrologer who strove to show that the earth revolves, not the heavens or the firmament, the sun and the moon. . . . This fool wishes to reverse the entire science of astronomy; but sacred Scripture tells us [Joshua 10:13] that Joshua commanded the sun to stand still, and not the earth.[8]

Not to be outdone, Luther's close collaborator and friend Melanchthon wrote, in 1549:

> The eyes are witnesses that the heavens revolve in the space of
> twenty-four hours. But certain men, either from the love of novelty,
> or to make a display of ingenuity, have concluded that the earth
> moves; and they maintain that neither the eighth sphere [the celestial
> sphere] nor the sun revolves. . . . Now, it is a want of honesty and
> decency to assert such notions publicly, and the example is pernicious.
> It is the part of a good mind to accept the truth as revealed by God
> and to acquiesce in it.[9]

Calvin also sermonized:

> Who will say that the Sun does not move and that it is the Earth that
> shifts and turns.[10]

Hence, by the mid-sixteenth century, both the Church of Rome and
the leaders of the Protestant movements agreed, though for different
reasons, that Copernicus's ideas were wrong. With this array of religious leaders lined up against heliocentrism for theological reasons,
few astronomers were eager to challenge geocentrism publicly.

Giordano Bruno, born in 1548 c.e. in Nola, Italy, and one of
the most radical and imaginative thinkers of the late sixteenth
century, was among the few who spoke freely in agreement with
Copernicus. Among his many other transgressions, Bruno argued
that God and the world are one, that the universe was infinite, that
God may have made many other inhabited worlds, that the world
has a soul, and that Christ was not God but rather a skilled magician.
Bruno was charged in Rome as a heretic in 1576, excommunicated
by the Calvinist Council in Geneva in 1581, lost his lecturing privi-
leges at Anglican Oxford in 1584, excommunicated by the Luther-
ans in Germany in 1587, convicted by Venetian Inquisitors in 1591,
extradited into the hands of the Roman Inquisition in 1592, and
finally, after being imprisoned for eight years, condemned and
burned at the stake in Rome in 1600. While he almost certainly was
condemned for heresies, of which heliocentrism was not formally
one (yet), as one of heliocentrism's more visible spokespersons

Figure 3.3: Tycho Brahe at age forty, in 1586, from the frontispiece to Brahe's *Astronomical Letters*, published in 1596.

Bruno probably did not help the cause of heliocentrism among Catholics or Protestants.

The most important astronomer of the latter half of the sixteenth century was Tycho Brahe (1546–1601 c.e.), a member of the Danish nobility who forsook the court for the scientific life. Tycho made several important discoveries that undermined Aristotle's cosmology. First, he discovered a nova, a new star, in 1572. This star appeared where none had been before and then, after a few months during which it underwent several changes of color, disappeared altogether. In the Aristotelian aether, stars did not and could not change in color or brightness or in any other quality; hence, Tycho's nova was an object considered by all at the time to be a phenomenon occurring in the Earth's atmosphere. Tycho, however, attempted to measure the distance to the nova and found that it was immeasurably distant, placing it well beyond the moon and therefore well outside of the Earth's atmosphere. Consequently, Tycho's measurements of the new star revealed that the perfection of Aristotle's heavens was a fallacy. Second, Tycho studied a comet in 1577. His calculations from his observations produced another visible challenge to Aristotle's universe; Tycho was able to prove that this comet, as well as six others he would observe over the next two decades, was much farther away than the Moon, whereas in an Aristotelian universe comets were luminous vapors in the Earth's atmosphere.

By the time Tycho passed from the scene, the Aristotelian universe was untenable to natural philosophers, as scientists were then called. The comets had shattered the celestial spheres and the rational arguments that had created the aether were shown false; thus, Aristotle's physics was now on a shaky foundation. Finally, Tycho's observations forced Tycho himself to abandon the Ptolemaic universe in favor of a partly heliocentric, partly geocentric "Tychonic" universe. In the universe according to Tycho, the Moon and Sun orbited the Earth while Mercury, Venus, Mars, Jupiter, and Saturn orbited the Sun. By virtue of having five planets orbit the Sun, the Sun carried those objects with it as it orbited the Earth. In this way,

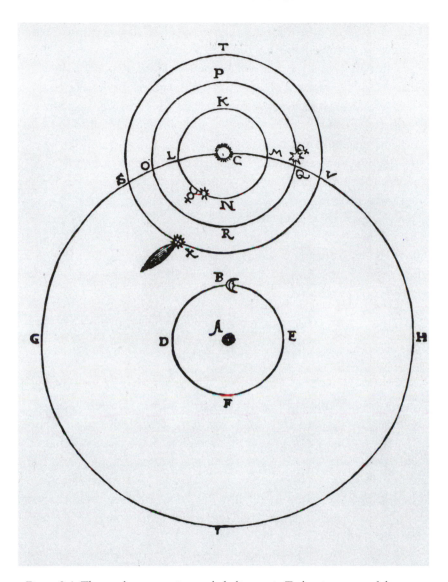

Figure 3.4: The partly geocentric, partly heliocentric Tychonic system of the world. The orbits of three planets are shown, centered on the Sun. The celestial sphere and the orbits of the Moon and Sun are centered on the Earth. As the Sun orbits the Earth, it carries the heliocentric planetary orbits with it. The comet of 1577 is shown at the position marked with the X. From Tycho Brahe's *De Mundo aetherei recentioribus Phaenomensis* of 1588. (Image courtesy of Image Select / Art Resource, NY)

Tycho preserved a vestige of geocentrism since all objects, one way or another, revolved about a geostatic Earth.

Ultimately, the most important work of Tycho was his thirty-eight years of systematic observations of the positions of the planets, obtained during the years 1563 through 1601. In his instrument workshop, he and his master craftsmen designed and built dozens of instruments, including new designs for sextants and armillaries as well as enormous rulers, all of which enabled him to make measurements of stellar and planetary positions with an accuracy at least eight times better than anyone previous to him had achieved. Tycho's observational records would become his legacy, bequeathed to his protégé, the mathematician Johannes Kepler, whom Tycho hired in 1600, shortly before his own death. Kepler, with Tycho's data in his possession, would become midwife to the birth of the Copernican Revolution.

Kepler was born in Weil der Stadt, in southern Germany, in 1571. In contrast to Tycho's noble birth, Kepler's father was a ne'er-do-well mercenary soldier, while his mother, the quarrelsome daughter of an innkeeper, was tried for witchcraft over a four-year period from 1617 to 1620. A poor boy, Kepler earned his M.A. from the University of Tübingen as a scholarship student. He then became a mathematics teacher in a Lutheran high school in Graz. When Roman Catholic authorities took over Graz, Kepler was ordered by Counter-Reformation vigilantes to leave the area. Kepler landed a new position as Tycho's assistant, in Prague, and soon thereafter, when Tycho died, inherited all of Tycho's observing records and became the court mathematician and astrologer to the Holy Roman Emperor Rudolph II.

Kepler is best remembered as the discoverer of three laws of planetary motion, and indeed, these are the aspects of Kepler's life's work on which we need to focus. These three laws are:

1. planets orbit the sun in elliptical paths with the Sun at one focus of the ellipse;
2. the straight line joining the Sun and a planet sweeps out equal areas in equal times;

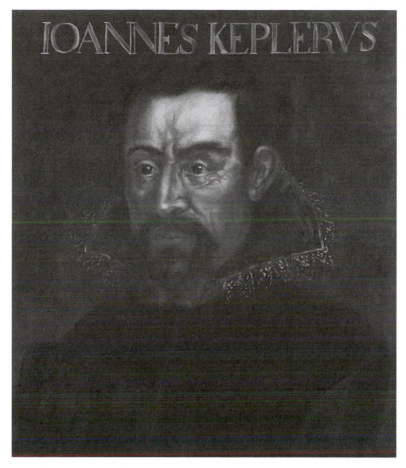

Figure 3.5: Portrait of Johannes Kepler. (Reproduced with permission of Istituto e Museo di Storia dell Scienza iconographic archive, Florence)

3. the squares of the periods of the revolution of the planets are directly proportional to the cubes of the semi-major axes of their orbits.

Simply put, the first law asserts that planets orbit the Sun, in support of the Copernican heliocentric model, but that they do not orbit in circular paths. Instead they follow elliptical paths. Roughly speaking, ellipses are flattened circles. Whereas a circle has a center and a radius, an ellipse is characterized by a short, or minor, axis and

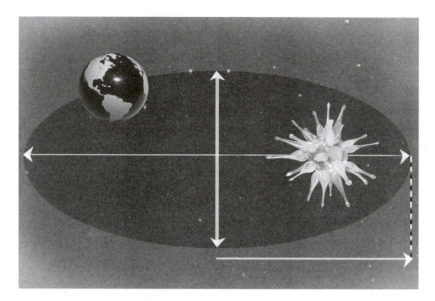

Figure 3.6: Kepler's first law: planets orbit the Sun in elliptical—not circular—paths. The Sun is at one focus of the ellipse; nothing is at the other focus. Half the length of the long axis of the ellipse, shown by the arrow just below the ellipse, is called the *semi-major axis*. The position along the elliptical orbit that is closest to the Sun (at the right edge of the ellipse) is the perihelion; the position that is farthest from the Sun (at the left edge of the ellipse) is the aphelion. (Artwork by Adrienne Outlaw ©)

a long, or major, axis (half of the major axis, known as the semi-major axis, is an important characteristic of an ellipse and of orbits) and two foci, which are found equidistant from the center and along the major axis; the separation of these two foci and the size of the major axis control the shape of the ellipse. A circle is a special case of an ellipse for which the two foci are located at the same point, in which case the major and minor axes become equal.

What Kepler has done through this simple statement, which took him five years of intense labor to discover, is demonstrate that one of the most fundamental tenets of Aristotelianism, perfect circular motion, which inspired and controlled the thinking of every astronomer from Hipparcus to Copernicus, is absolutely wrong. Planets, whether they move in geocentric (the Moon) or heliocentric orbits, do not move in circles.

Kepler's second law asserts that a planet in its elliptical orbit moves at a higher angular velocity when close to the Sun than when far from the Sun. Specifically, the mathematical product of the planet's angular velocity and the square of its distance from the Sun cannot change; this mathematical product is directly proportional to the quantity modern physicists refer to as *angular momentum*. When a physical quantity, such as angular momentum, cannot change, physicists say that this quantity is "conserved." Conserved quantities, or conservation laws, are among the most important concepts in modern physics, and Kepler had discovered the first conservation law. As a result of this law, we know that when a planet moves in its orbit from a first position close to the Sun to a second that is twice as far from the Sun as the first, the angular velocity decreases by a factor of four.

Kepler's third law is a very powerful tool, in that it enables an astronomer to determine the distance of a planet from the Sun without directly measuring the distance. How? One simply measures the time required for the planet to orbit the Sun once. With this measurement of the orbital period, Kepler's third law immediately yields the size of the orbit of a planet (the semi-major axis). While to Kepler, this was an elegant but unexplainable piece of mathematics, within a century Isaac Newton would show that the relationship is an unavoidable consequence of the law of gravity.

For one very critical reason, Kepler's mathematical laws would have been the envy of Hipparcus, Ptolemy, Copernicus, Tycho, and every other astronomer who ever preceded him: they work. Using Kepler's laws, anyone mathematically inclined can accurately calculate the future position of any of the planets. Kepler has satisfied Hipparcus's requirements for mathematical accuracy, though he has done so via a heliocentric universe in which he has abandoned Aristotle and the idea of circular motion.

Now how many planets do we have, and why is one object a planet while another is not? To the ancients, a planet was a wandering star, of which they knew of seven: Mercury, Venus, the Moon, the Sun, Mars, Jupiter, and Saturn. Then, over nearly two thousand years, the ancient Greek philosophers and the medieval Christian scholars, as well as many in between, built an enormous edifice to support, justify, prove, and proclaim the physical and biblical necessity of the existence of these, and only these, seven planets. Suddenly, with Kepler's mathematics demonstrating that the universe is heliocentric, not geocentric, the ancient definition of a planet made no sense. For earthbound observers, the Earth hardly can be seen as a wandering star while at least one object previously considered a wandering star, the Sun, was now considered fixed and motionless near the center of the heavens, even though it appears to wander through the fixed stars. Apparently appearances can be deceiving. We now can dispense with the Old Astronomy definition in the *Oxford English Dictionary*.

Having invalidated the old definition for a planet, we need a new one. Kepler has effectively provided us with a new definition: *a planet is an object that orbits our Sun in accord with his three laws.* And in 1619, the year Kepler published the third of his three laws, astronomers knew of six such planets: Mercury, Venus, Earth, Mars, Jupiter, and Saturn. Notably, this definition excludes the Moon. What, then, is the Moon? It does obey Kepler's laws; however, it orbits the Earth, not the Sun, with the Earth at one focus of the Moon's elliptical orbit.

As we shall see, by 1610, before Kepler's third law was published, before any of his laws were widely known or generally accepted, and many decades before the heliocentric hypothesis of Copernicus had firmly displaced geocentrism, Galileo would discover four more alleged planets. Rather than drop from the seven wandering stars to the six heliocentric planets, Galileo will up the ante.

4
Sixteen Planets

O

One of the fathers of modern science, along with Johannes Kepler and William Gilbert, the latter of whom published his treatise *De Magnete* in 1600, and of the scientific method and the defendant in the trial of the millennium in 1633, Galileo Galilei was born in 1564, only twenty-one years after the death of Copernicus. Though raised as a devout Catholic in Florence, Italy, he apparently inherited from his father, a musician and artist, a habit of questioning authority. His attitude did not endear him to his medical school professors at the University of Pisa, as Galileo questioned the correctness of Galenic medicine. Thus, Galileo found himself drawn toward mathematics, a field in which correctness was easier to check and challenge than in other fields.

By the late 1590s, Galileo was in correspondence with another continental mathematician, Kepler, and apparently they both became convinced of the correctness of Copernicus's view of the universe about the same time, around 1595. But while Kepler went public with his support of Copernicus in his 1595 book *Mysterium Cosmographicum* (Mystery of the Universe) and devoted his life to the discovery of mathematical formulae that accurately describe heliocentric orbits, Galileo was a private Copernican, spending most of the next fifteen years doing important physics experiments in mechanics and optics.

In 1609, the same year in which Kepler published the first two of the laws that would bear his name, Galileo learned about a newly

Figure 4.1: Portrait of Galileo Galilei. (Reproduced with permission of Istituto e Museo di Storia dell Scienza iconographic archive, Florence)

invented optical device that, using two specially designed lenses, made distant objects look bigger. Galileo, being an inventor and tinkerer, began grinding his own lenses from high-quality Venetian glass and quickly made his own spyglass, with a magnifying power of three. Such an instrument makes objects look three times closer. If the object under observation is a circle, and if the measured radius of a circle increases by a factor of three because the object appears three times closer, then the area of the circle (i.e., the disklike image

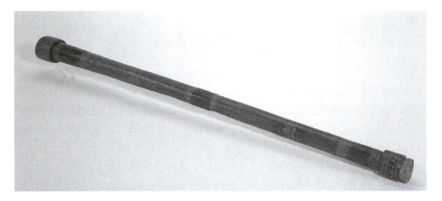

Figure 4.2: Galileo's telescope. (Reproduced with permission of Istituto e Museo di Storia dell Scienza iconographic archive, Florence)

of a planet), which increases in proportion to the area of the circle (or the square of the radius), will be nine times bigger.

Galileo's next device was even better, with a magnifying power of twenty. On the twenty-ninth of August 1609, Galileo, who was then Professor of Mathematics at the University of Padua in the Republic of Venice, led a procession of Venetian merchants out into the Piazza San Marco, where he demonstrated how his spyglass could be used to spot distant, incoming merchant vessels still at least two hours from port. Such a device was quickly seen by the merchants as an important money-saving device for their businesses and of obvious military value. As a result, Galileo was commissioned to produce a great many spyglasses for the city, and his salary was doubled. Shortly before one merchant ship arrived from Holland with a load of less expensive, Lippershey-crafted spyglasses, Galileo leveraged his new fame and prestige into a position as the court mathematician to the Medici family in Florence.

At some point, once in Florence, Galileo turned this device toward the nighttime sky and quickly made a number of discoveries that brought him immediate and widespread fame. In 1610, he published his first discoveries in a pamphlet entitled *Sidereus Nuncius* (*Sidereal Messenger*). Among these discoveries, he asserted that the Moon was like the Earth. The Moon, he reported, was not a perfect sphere of aether, as claimed by the Aristotelians; rather, it had mountains and valleys and a highly irregular surface, with constantly changing patterns of shadows. Three years later he would make a similar discovery about the Sun. He would observe dark spots on the surface of the Sun. These spots changed in size and shape and moved in a way consistent with them being part of a rotating Sun; thus, the Sun, like the Moon, was imperfect and not made of aether.

FOUR NEW PLANETS: THE MEDICEAN SATELLITES

Galileo's most profound discovery was identifying four small, starlike points of light that moved around the sky with Jupiter. In fact, after watching Jupiter and its four attendants for only eight days, Galileo concluded that these four objects were in orbit around Jupiter. He wrote of his discovery of "four planets flying around the star of Jupiter at unequal intervals and periods with wonderful swiftness; which, unknown by anyone until this day, the first author detected recently."[1] Knowing from whence came his political patronage, Galileo took the opportunity to name these four newly discovered planets the Medicean stars.

Jupiter, Galileo realized, is a center of motion like the Earth; the motion of the Medicean stars around Jupiter demonstrates that Aristotle was wrong about the Earth being the only center of motion in the universe. And if both Earth and Jupiter can be centers of motion, then surely the Sun also can be one. Consequently, Galileo became convinced that his discoveries of the Medicean stars proved Copernicus right. Over the next twenty-three years, Galileo would single-handedly and single-mindedly stand up to two popes and the Inquisition, desperately but unsuccessfully trying to convince the powerful leaders of the Roman Catholic Church to accept Copernican heliocentrism and reject Aristotelian and Ptolemaic geocentrism. After two battles with the Inquisitors, in 1616 and again in 1633, Galileo was forced to recant his heliocentric opinions and was placed under house arrest for the remaining nine years of his life.

But what about Galileo's so-called Medicean stars? They cannot be placed under house arrest nor made to disappear again, to live

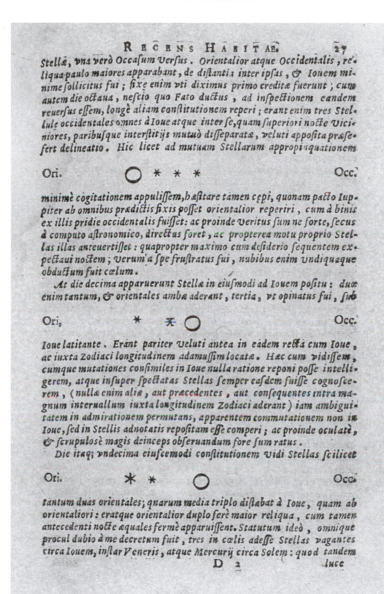

Stellæ, vna verò Occasum versus . Orientalior atque Occidentalis , re-
liquæ paulo maiores apparabant, de distantia inter ipsas , & Iouem mi-
nimè sollicitus fui ; fixæ enim vti diximus primo creditæ fuerunt ; cum
autem die octaua , nescio quo Fato ductus , ad inspectionem eandem
reuersus essem, longè aliam constitutionem reperi ; erant enim tres Stel-
lulæ occidentales omnes à Ioue atque inter se, quam superiori nocte vici-
niores, paribusque interstitijs mutuò disseparatæ , veluti apposita præse-
fert delineatio . Hic licet ad mutuam Stellarum appropinquationem

Ori.   ○ ＊ ＊ ＊   Occ.

minimè cogitationem appulissem, hæsitare tamen cepi, quonam pacto Iup-
piter ab omnibus prædictis fixis posset orientalior reperiri , cum à binis
ex illis pridie occidentalis fuisset: ac proinde veritus sum ne forte, secus
à computo astronomico, directus foret , ac propterea motu proprio Stel-
las illas anteuertisset : quapropter maximo cum desiderio sequentem ex-
pectaui noctem ; verum à spe frustratus fui , nubibus enim vndiquaque
obductum fuit cœlum .

At die decima apparuerunt Stellæ in eiusmodi ad Iouem positu : duæ
enim tantum, & orientales ambæ aderant , tertia , vt opinatus fui , sub

Ori,   ＊   ＊○   Occ.

Ioue latitante . Erant pariter veluti antea in eadem rectâ cum Ioue ,
ac iuxta Zodiaci longitudinem adamussim locatæ . Hæc cum vidissem ,
cumque mutationes consimiles in Ioue nulla ratione reponi posse intelli-
gerem, atque insuper spectatas Stellas semper easdem fuisse cognosce-
rem , (nulla enim alia , aut præcedentes , aut consequentes intra ma-
gnum interuallum iuxta longitudinem Zodiaci aderant) iam ambigui-
tatem in admirationem permutans, apparentem commutationem non in
Ioue, sed in Stellis adnotatis repositam esse comperi ; ac proinde oculatè ,
& scrupulosè magis deinceps obseruandum fore sum ratus .

Die itaq; vndecima eiuscemodi constitutionem vidi Stellas scilicet

Ori.   ＊ ＊   ○   Occ.

tantum duas orientales; quarum media triplo distabat à Ioue , quam ab
orientaliori : eratque orientalior duplo ferè maior reliqua , cum tamen
antecedenti nocte æquales fermè apparuissent. Statutum ideò , omnique
procul dubio à me decretum fuit , tres in cœlis adesse Stellas vagantes
circa Iouem, instar Veneris, atque Mercurij circa Solem : quod tandem

D 2   *luce*

Figure 4.3: Galileo's sketches of the moons of Jupiter, from *Sidereus Nuncius*. (Reproduced with permission of Istituto e Museo di Storia dell Scienza iconographic archive, Florence)

Figure 4.4: The four Galilean moons. *From left to right* (and by increasing orbital distance from Jupiter): Io, Europa, Ganymede, and Callisto. (Image courtesy of NASA/JPL-Caltech)

outside of human knowledge and memory. In fact, these spots of light that followed Jupiter were seen even by many; most important, the Jesuit astronomers at the Collegio Romano, led by the foremost astronomer of the time, Christopher Clavius, confirmed Galileo's observations in 1610.

While Galileo called them "Medicean stars," they quickly became known as planets. Counting these four new planets, in 1610, by Galileo's reckoning we must have eleven rather than seven planets. If we accept the Copernican view of the universe, six of these planets orbit the Sun while five of them orbit two of the heliocentric planets. In total then, we have six heliocentric planets, one geocentric planet, and four Jovian-centric planets. All eleven of these planets obey a form of Kepler's laws, whereby we simply replace "with the Sun at one focus" of their elliptical orbits to "with the Sun, Earth, or Jupiter at one focus," choosing the focal object appropriately to the situation.

How, then, should we classify objects? Are all eleven objects planets? Do three subcategories of planets exist, depending on what object is at one focus of the elliptical orbit? Or do we have different types of objects, with one type orbiting the Sun and a second type orbiting a planet?

Soon after Galileo published his *Sidereal Messenger*, Kepler published a response, in which he used both *planet* and *moon* interchangeably to describe Galileo's discoveries. Kepler was convinced that life probably existed on all planets, which to him were objects orbiting stars. In contrast, he didn't feel compelled to infer the existence of life on lowly objects orbiting planets:

> If you had discovered any planets revolving around one of the fixed stars, there would now be waiting for me chains and a prison among Bruno's innumerabilities. . . . Therefore, by reporting these four planets revolve, not around one of the fixed stars, but around the planet Jupiter, you have for the present freed me from the great fear that gripped me as soon as I had heard about your book. . . .
>
> These four little moons exist for Jupiter, not for us. Each planet in turn, together with its occupants, is served by its own satellites.[2]

On the one hand, it might appear that Kepler thinks of moons as planets; on the other hand, Kepler is perhaps the first to distinguish between the general concept of a "moon" as a satellite of a planet and "planet" as a satellite of a star. Note that for Kepler, the existence of moons around Jupiter is a demonstration that living beings inhabit Jupiter, since the purpose of the moons is to provide light and beauty to the inhabitants of that planet.

In 1633, Galileo was put on trial by the Inquisition authorities in Rome. Galileo had already been admonished, in 1616 by Cardinal Robert Bellarmine, the Consultor of the Holy Office and Master of Controversial Questions, that Copernicanism could not be "defended or held."[3] When he was actually brought before the Inquisitors in 1633, Galileo, and not Copernicanism, was on trial. Galileo was convicted of violating the restrictions promulgated by Cardinal Bellarmine in 1616, of having held and defended Copernican ideas. The sentence announcing Galileo's guilt asserted that Galileo had held two opinions that were considered formally heretical, "that the Sun is the center of the world and motionless" and "that the Earth is neither the center of the world nor motionless but moves." Galileo was pronounced "vehemently suspected of heresy" and condemned to "formal imprisonment in this Holy Office."[4] In legal terms in 1633, the words "suspected of heresy" were merely a less heinous form of heresy than "formal" heresy; they did not imply suspicion with the absence of proof, as is connoted by modern usage of the word "suspected." Thus, by virtue of believing that Copernicus was right, Galileo was a heretic. Anyone of less stature than Galileo, i.e., virtually everyone, almost certainly would have suffered a much harsher fate than mere imprisonment.

Shortly after his trial, Galileo was released to spend the last eight years of his life under house arrest in Florence. The harsh treatment of Galileo sent a chill through the European scientific community; in fact, in 1633 René Descartes chose to abandon work on his manuscript *Le Monde*, in which he attempted to provide a new physical theory of the universe and on which he had worked from 1629 to 1633. Decartes died in 1650; his unfinished manuscript *Le Monde* was only published posthumously, in 1664. Yet despite the harsh treatment of Galileo, in scientific circles the geocentric universe was dead: within a few decades, the paradigm shift to the heliocentric universe was complete. By midcentury, all astronomers agreed that the Earth was indeed a planet and the universe had eleven planets and a central Sun. Six of these planets revolved around the Sun, one revolved around Earth, and four revolved around Jupiter.

SATURNIAN SATELLITES

In 1655, Christiaan Huygens (1622–1695 C.E.) and his brother built a 12-foot-long telescope with a 2.5-inch-diameter lens. With this telescope Huygens discovered a satellite, Titan, in a sixteen-day orbit around Saturn. In 1659, after dedicating four years to studying the puzzling appendages of Saturn first described by Galileo, Huygens wrote, in *Systema Saturnium* (The System of Saturn), that "Saturn is surrounded by a thin flat ring not touching it anywhere."[5]

Jean-Dominique Cassini, working at the Observatory of Paris from 1671 to 1684, discovered four more satellites of Saturn.[6] Including the five Saturnian satellites discovered by Huygens and Cassini, all of which were considered planets at the time, the total number of objects called planets had grown to sixteen by 1684.

Figure 4.5: Several of Saturn's icy moons are visible as tiny starlike objects in or near the ring plane above and to the left of Saturn, in this Hubble Space Telescope image from 1996. These moons are *from left to right*, Enceladus, Tethys, Dione, and Mimas. (Image courtesy of NASA/JPL-Caltech)

Were all sixteen of these objects planets? Formally, only six of them obeyed Kepler's laws, since ten of them did not revolve about the Sun. Does this provide a physical basis for distinguishing planets from other objects? Kepler apparently thought so. The other ten, however, did orbit in elliptical paths and sweep out equal areas in equal times, and for the four objects in orbit around Jupiter and the five in orbit around Saturn, the squares of their orbital periods were proportional to the cubes of their semi-major axes. These other ten objects just happened to orbit the Earth, Jupiter, and Saturn rather than the Sun. Astronomers and writers at the close of the seventeenth century were not as quick to dismiss these nonheliocentric objects as not planets as Kepler probably would have been.

Bernard le Bovier de Fontenelle (1657–1757 c.e.), a major figure of the French Enlightenment, wrote one of the most popular books of the seventeenth century, *Entretiens sur la pluralite des mondes* (*Conversations on the Plurality of Worlds*). This book, published in 1686 and one of the first great popularizations of science, had been translated into English by 1688 and into nine other European languages by 1800.[7] *Pluralite des mondes*, written in a popular fashion as a dialogue between the author and a beautiful noblewoman, by and large put forth the thesis that the many other worlds in the universe are inhabited; in fact, the greatness of God necessitated that they all must be inhabited. In this dialogue, we find the following inter-change, as the author explains how the planets are now known to orbit the Sun rather than the Earth:

> [Fontenelle, as the author]. "And as the earth moves in the circle it describes round the sun, the moon follows the earth in turning round it. And if the moon do move round the sun, it is only because she will not abandon the earth."
> "I understand you," said she. "I love the moon for staying with us when all the other planets have left us."

On the second night of the conversation imagined in this dialogue, the author tells Madam the Marchioness:

> "Well," said I, "since the sun . . . is now immovable, and no longer a planet, and that the earth that moves round the sun is now one, be not surprised if I tell you, the moon is another earth, and is by all appearances inhabited."

Later Fontenelle discusses other similarities between the Earth and Moon:

> "But the other planets shine only as being enlightened by the sun. The sun communicates his light to the moon, which reflects it back upon the earth, as the earth, without doubt, reflects it back again to the moon."[8]

Ultimately, Fontenelle argues that despite the presence of dark spots on the Moon, believed by many to be seas and named *mare*, the Moon was waterless and probably lifeless. Clearly, Fontenelle did not distinguish the Moon as a different kind of object from the Earth except in the inability of the Moon to support life.

In this claim that the Moon was lifeless, Fontenelle received the support of Huygens who noted, in his *Cosmotheoros* (*Celestial Worlds Disover'd*), published posthumously in 1698, that a century of telescopic observations served to disprove the claims that the Moon had seas or an atmosphere. The Moon, being close enough for careful, systematic study, was gradually giving up her secrets. The mare were rough cavities, no rivers had been observed on the lunar surface, and the crisp lunar edge and absence of any evidence of clouds demonstrated the absence of air. Therefore, the Moon must be devoid of life.

Huygens then spent considerable time in *Cosmotheoros* describing what was known about each planet and each planetary satellite. In truth, there was little he could say because, despite the efforts of astronomers and telescope builders of the seventeenth century, all the celestial objects except the Moon were simply too distant to reveal much. The telescopes showed most clearly that distant planets and their satellites all appeared as disks, not pinpoints of light like the stars. Thus, astronomers could conclude that all the known planets and satellites were spherical. In addition, some information about colors and color patterns and rotation rates could be discerned. But otherwise, all save the Moon hid their secrets well from Huygens and his contemporaries. Nevertheless, perhaps on the basis of his efforts to show how different from the Earth, and how lifeless, the Moon is, Huygens described the satellites of Jupiter and Saturn as "secondary planets" and argued that Earth's Moon was representative of this group, whereas the Earth itself was representative of the primary planets.[9]

At least to Huygens, at the end of the seventeenth century, the presence of running water, oceans, an atmosphere, and life served to distinguish a primary planet from a secondary planet. Life could and should exist on primary planets; life could not exist on secondary planets. Thus, if this view held, the number of known primary planets—life-bearing satellites of the Sun—would drop back to six, while the newly defined classification group of secondary planets—lifeless satellites of planets—would be populated by at least ten objects.

While Cassini and Huygens were busy building telescopes and discovering the moons of Saturn, and while they and many others were occupied with speculations about extraterrestrial life on the Moon and beyond, dramatic changes were occurring elsewhere in the physical and astronomical sciences that would have profound effects on our view of the cosmos and on our understanding of the types of objects present in the universe. Aristotle's luminous vapors, first lifted by Tycho into heliocentric orbit, are about to be lofted upward much further by Isaac Newton and Edmund Halley.

5

Not Everything That Orbits the Sun Is a Planet

O

Comets do not look or behave like any other objects seen in the sky. Unlike shooting stars that hurtle across the sky in a fraction of a second, comets spend weeks or months slowly inching across the sky, wandering slowly through the constellations like planets. Comets linger, changing their appearances slowly, patiently.

Our word for comets, which comes from the Greek *kometes*, meaning "long haired (star)," illustrates just how unlike stars and unlike planets such objects are. To ancient peoples who experienced a much closer relationship with the sky than moderns, comets lingering in the sky for many months would have been phenomenal celestial events to witness, and the ancients would have seen comets quite often. So-called naked-eye comets appear with regularity, about every two or three years. A few of these, like Comet Halley in 1910, Comet West in 1976, and Comet Hyakutake in 1996, are great comets, with tails extending across half the sky. Such objects could not have escaped the notice of the astronomers of ancient civilizations, and in cultures that developed in which events in the sky had religious or political significance, comets quickly took on great importance.

Aristotle called comets "luminous vapors," and that indeed describes their appearance well. A comet normally is bright and compact at one end, called the head, with a long, diffuse tail (or tails) extending many degrees across the sky from the head. When a comet

Figure 5.1: Comet Hyakutake (1996). (Picture courtesy of Rick Scott and Joe Orman)

first appears, assuming an observer notices the comet at the earliest possible time, only the head is detectable, and the comet looks like a planet, as this starlike object wanders past the fixed stars over nights and weeks. Gradually, the comet head grows some whiskers, usually extending in only one direction; then those whiskers grow longer and longer, becoming the cometary tail. At long last, the tail

Figure 5.2: Comet Hale-Bopp (1998). (Picture courtesy of Rick Scott)

shortens and fades away; finally, the head of the comet fades and disappears from view.

What might the ancients have thought of such phenomena? If nothing else, they paid attention. Chinese astronomical records include definitive observations of Halley's comet as far back as 240 B.C.E, with a possible sighting as long ago as 1057 B.C.E[1] As far back as two thousand years ago, Chinese astronomers recorded on average about twenty comets per century. Babylonian records suggest possible observations of comets going back as far as 3000 B.C.E, though not with the kind of accuracy as the Chinese records. The Chaldeans, in ancient Babylonia, built observing towers, or ziggurats, in almost all their cities. From these towers, they recorded their observations of events in the sky, including aurora, comets, unusual cloud patterns, and appearances of planets and stars. Presumably, the close attention paid to the sky in these ancient cultures corresponded closely to the importance placed on celestial happenings.

Whatever the meaning attributed to the appearance of a comet, our interest is in what comets were understood to be. Among the pre-Socratic Greek philosophers, in the sixth century B.C.E Xeno-phanes believed that the Sun, Moon, planets, and stars were all fiery vapors exhaled by the Earth each day and burned out a half day later. According to Xenophanes, comets, though they did not appear with the regularity of the other objects, were also fiery atmospheric phenomena emerging from exhalations of the Earth. A century later, Anaxagoras suggested that comets emerged from the conjunction of two planets, while Diogenes of Apollonia (c. 430 B.C.E) claimed that comets were chains of stars (and that stars were fiery rocks). Meanwhile, Hippocrates of Chios (c. 430 B.C.E) suggested that comets were "rare" appearances of planets, while some members of the Pythagorean school believed that comets were not even real, that they were optical illusions resulting from reflected sunlight.[2]

By the time of Aristotle, a half dozen or more competing theories of comets had emerged. The Roman philosopher and statesman Seneca the younger (4 B.C.E–65 C.E.), writing in his metaphysical treatise *Quaestiones Naturales*, wrote about two competing theories from earlier Greek times, both of which were reported by the Greek authors to have come from the Chaldean astronomers. The first theory described comets as violent, fiery whirlwinds in the atmosphere. The second explanation credited the Chaldeans with the idea that comets were a kind of wandering star (i.e., a planet with a known orbit).[3] These two hypotheses likely depict well both the wealth of knowledge and understanding as well as the ignorance and confusion that characterized ancient understandings of comets.

In his *Meteorology*, Aristotle argued that comets were luminous vapors, perhaps some kind of interaction between the basic substances of fire and water and air. For Aristotle, in order to be consistent with the rest of his natural philosophy, comets necessarily were sublunary, or atmospheric, phenomena since they were objects that continually changed in appearance. Aristotle's ideas did have competition, but as we also have seen, Aristotle's opinions, once established, became entrenched and held sway for nearly two thousand years. Since comets were not thought of by Aristotle as wandering stars, astronomers and natural philosophers all agreed, during the two millennia preceding the work of Tycho Brahe, that seven and only seven planets graced the celestial spheres.

TYCHO'S COMET

On November 13, 1577, while out fishing in the evening, Tycho noticed a bright star in the western sky. Tycho, at that time the best observational astronomer in Europe and probably on the Earth, knew immediately that this object was not a planet. Tycho knew exactly the positions in the sky of all the planets and all the stars visible to the naked eye. He understood, immediately, that the position of this object did not correspond with that of any known planet or star. By the time darkness had set in, he could see that his "star" had a tail and must be a comet. Tycho, being a devout Aristotelian at this time, assumed that comets were atmospheric vapors. Being the great astronomer that he was, he knew he could measure the distance to this comet, since he was capable of measuring the distance to any object closer than the moon. As no astronomer had ever measured the distance to a comet, and because Tycho believed this distance would give him the height of the comet in the atmosphere, such a measurement would offer a new and unique contribution to astronomy.

To measure distances, Tycho measured the apparent position of the comet with respect to the very distant stars at different times throughout the evening. In other words, he measured the parallax of the comet. Initially, in December, Tycho was convinced that his comet showed no parallax, which would require that it be much more distant than the Moon. After two months of measurements, after which the comet had faded from view, his more refined calculations from his measurements placed an upper limit on the parallax of the comet that was at least four times smaller than the parallax of the Moon. In other words, the comet had to be at least four times more distant than the Moon. Tycho's measurements of parallax therefore placed the comet of 1577 in the celestial realm, well outside of the atmosphere. Tycho observed other comets in 1580, 1582, 1585, 1590, 1593, and 1596, always getting similar results: all of the comets were at least three times farther away than the Moon. Aristotle was wrong. Change was happening in the heavens. In addition, from his many observations of each comet, Tycho attempted to calculate geocentric orbits for his comets but failed. He concluded that comets *did not orbit the Earth* and probably orbited the Sun.

By no means did the astronomical community immediately embrace the idea that comets are heliocentric rather than geocentric. The great Galileo, defender of the Copernican theory, argued that comets were illusions present in the atmosphere and that Tycho's measurements of cometary distances were meaningless, since they were measurements of reflections or refractions in the atmosphere of some poorly understood but local event. On the other hand, Tycho's successor Kepler defended the Tychonic hypothesis that comets are celestial objects; he even understood the nature of cometary tails— that they always point away from the Sun and that this must be due to some kind of interaction with radiation from the Sun—nearly 350 years before modern astronomers worked out the details.

Nevertheless, in trying to understand the meaning of planets, we now have the enticing and intriguing possibility of a new phenomenon: celestial objects whose mechanical properties (they do not orbit the Earth; they move through the fixed stars) are crudely identical to those of the planets but whose behavior otherwise unequivocally distinguishes them from planets. Half a century later, as heliocentrism displaced geocentrism in the post-Galileo era, comets would take their place in the celestial zoo alongside the Sun, the stars, the Moon, the moons of other planets, and the planets themselves. Most important, comets provided evidence of celestial objects that did not orbit the Earth and yet were not planets. However, comets were distinguished from planets in that planets had well-known, well-understood, predictable orbits, whereas comets appeared once and then disappeared forever. They were not thought to follow Kepler's laws and were thought, for most of the next century, to move in straight paths through the heavens, perhaps coming from beyond the planetary system and returning thereto.

EDMOND HALLEY'S PREDICTION

Edmond Halley (1656–1742 C.E.) studied his first comet at age twenty-three. While on vacation in Paris, Jean-Dominique Cassini let him observe the comet of 1680 at the Observatory of Paris. From his observations, Halley attempted to calculate an orbit for the comet, but like his predecessors and contemporaries, his efforts to force this comet to travel in a straight line path through the solar system came to naught. Halley's passion for comets, however, had been ignited. Halley witnessed his next comet in the late summer of 1682, while others observed and recorded the changing position of this comet of 1682 over the course of a month, from August into September. Two decades later, Halley correctly determined the orbital path of this comet, and by doing so he unlocked a great secret about the solar system.

Perhaps the greatest single unanswered scientific question in the late seventeenth century was that of deriving Kepler's laws of planetary motion from basic physical principles, from the dynamics of motion. In January of 1684, at a meeting of the Royal Society, Halley discussed this problem with Robert Hooke and Christopher Wren, with Hooke claiming that all the laws of motion could be derived from a mathematical inverse square formula. An inverse square relation implies that one quantity (e.g., gravitational force) decreases in proportion to the square of another quantity (e.g., distance times distance). Thus, if the distance between two objects doubles, the force of attraction between those two objects decreases by a factor of four. Hooke, however, could not back up his claim with a mathematical demonstration, and Wren, perhaps hoping to inspire Hooke to make good on his claim, offered a prize of forty shillings to anyone who might prove this claim within two months. Hooke did not claim the prize.

Seven months later, Halley paid a visit to Isaac Newton at Cambridge. As retold by Abraham DeMoivre, supposedly as recalled by Newton, Halley

> asked him what he thought the Curve would be that would be described by the Planets supposing the force of attraction towards the Sun to be reciprocal to the square of their distance from it. Sr. Isaac replied immediately that it would be an Ellipsis.[4]

Halley immediately urged Newton to provide the proof in the form of his calculations, which Newton was unable to do, claiming he could not find the papers on which he had done the work. But Halley persisted in urging Newton to show the world these important results, and three years later Newton published—at Halley's expense—*Principia*, one of the most important scientific treatises ever written.

Through the mathematics of the *Principia*, Newton formulated the law of gravity and, therefrom, derived Kepler's laws. He further asserted that the same law controlled the motions of moons around planets and comets around the Sun. In the first edition of the *Principia*, he calculated a parabolic orbit for the comet of 1680, although he appeared to think of comets as following extremely large elliptical orbits, since an infinitely large ellipse would be indistinguishable from a parabola. In *Principia* he wrote, "I am out of my judgment, if they are not planets of a sort, revolving in orbits that return into themselves with a continual motion."[5] Newton has concluded that comets and planets follow the same law of gravity; comets, however, follow parabolic (or infinitely large elliptical) orbits while planets travel in smaller, nearly circular, elliptical orbits. Thus, to Newton these planets could be distinguished from comets through the shapes and sizes of their orbits.

Newton's law of gravity was the mathematical tool Halley needed to continue his own investigation of comets. Because extremely large, parabolic orbits are indistinguishable (given the accuracy of the observations) from large elliptical orbits, and because the calculation of a parabolic orbit is significantly simpler (the number of unknowns is reduced from six to five)—though not simple—than the calculation of an elliptical orbit, Halley set out to calculate parabolic orbits for comets. Over a decade, from 1695 to 1705, he calculated parabolic orbits for twenty-four comets. Halley found that three of these comets—those of 1531, 1607, and 1682—had orbits so similar that they had to be one and the same object. Most notably, all three had unusual retrograde orbits; that is, they orbited the Sun in the opposite direction as the planets. In addition, the shapes and orientations of their orbits, their closest approaches to the Sun, and the eighteen-degree tilt of their orbits to the Earth's orbital plane were far too alike to be a coincidence, especially when so few comets were known, and he could explain the slight differences between their orbits as caused by the gravitational effects of Jupiter. Halley concluded that these three comets, as well as the comets of 1305, 1380, and 1456, were the same object, returning periodically, every seventy-six years, in what must surely be a large elliptical orbit (he was right about the comet of 1456 but wrong about the other two). Thus, Halley was led to predict that this comet would return again in 1758.

HALLEY'S COMET RETURNS

On Christmas Day 1758, an amateur astronomer and farmer in Germany spotted a comet, the one we now know as Halley's comet. One month later, on January 21, 1759, the professional astronomer Charles Messier identified the comet. Calculations of the orbit of this comet verified that it was indeed on an orbit like the comets of 1531, 1607, and 1682. Thus, not only was Halley's prediction proven true, Newton's laws had also been vindicated through their predictive power. This was a new way of doing science: take a "law" of physics, make a prediction, and determine the robustness, the accuracy of the proposed law.

As for the number of planets, comets obey all of Kepler's laws. Therefore, if Kepler's laws define planets, then logically comets now must be reclassified as planets. Do we then suddenly have as many planets as there are comets?

Comets had been known to the ancients and had never been called planets. Besides being visibly different kinds of objects from planets, their orbits were so highly elliptical that they appeared to be on parabolic orbits, and some apparently had extremely long orbital periods (the comet of 1680 was erroneously thought to have an orbital period of 575 years). The newly accepted understanding of comets generated no confusion between cometary and planetary orbits and, of course, comets already had a name.

Thus, the discovery that comets orbit the Sun did not increase the number of known planets, but they did something else even more important for our understanding of planets. With the return, in 1758, of the comet of 1682, the solar system was now known to include objects that orbit the Sun but that are not planets. *We have a new rule: not everything that orbits the Sun is a planet.* With this new rule, we also have a new problem: some objects satisfy Kepler's laws yet are not planets. Therefore, saying that an object follows Kepler's laws is not sufficient for labeling that object as a planet. What, then, is a planet?

By 1758, nearly a century had passed since the 1684 discovery of a fifth moon of Saturn. Moons were now commonplace components of the solar system, and the passage of time had led to the gradual acceptance of the nomenclature of "moons," which replaced "secondary planets" for such objects. The emergence of "moons" as a separate category of objects from "planets" led to the gradual evolution of a new component of the accepted rule for planets: planets orbit the Sun while moons orbit planets. We might state this more carefully: while planets and moons all obey Kepler's laws, moons have primary orbits around planets while planets have primary orbits around the Sun.

Comets are not planets. Moons are not planets. Thus, with these two unwritten rules in place for defining objects as "not planets," by the middle of the eighteenth century the number of planets had become firmly established as six: Mercury, Venus, Earth, Mars, Jupiter, and Saturn. Yet, while the criteria for determining which objects *were not* planets were growing, the criteria for determining which objects *were* planets were no longer unassailably clear, as an object now could satisfy all the accepted rules for being a planet (i.e., Kepler's laws) yet still not be a planet.

6
Uranus!

O

Subsequent to the reappearance of Halley's comet on Christmas Day in 1758, Newton's law of gravity metamorphosed from an equation that could explain phenomena astronomers already knew about to a theory that could predict previously unknown things about the universe: Halley had predicted that comets are part of our solar system, in closed orbits around the Sun, not interlopers from interstellar space, and he was right.

Overnight, the power that mathematical physics could bring to bear in revealing heretofore secrets of the universe was manifest. If the mysteries of the universe now can be probed through mathematics, might not other, yet undiscovered mathematical laws exist? One might further surmise that those who had the wherewithal to discover these laws might leave their footprints in history, to be remembered as intellectual giants and mentioned in the same breath as Kepler, Newton, and Halley.

One of the great mysteries of the solar system is the spacing of the planets. Why are the inner planets Mercury, Venus, Earth, and Mars so close together while Mars, Jupiter, and Saturn are so far apart? Why is the distance between the orbits of Mars and Jupiter greater than twice the distance from the Sun to Mars? Are the spatial separations of the planets necessary or accidental?

A term that is useful for measuring relative sizes of orbits in the solar system, and which comes from the mathematical definition of an ellipse, is the *semi-major axis* (usually designated by the letter *a*), which is one-half of the long axis of an ellipse. The semi-major axis also is a fundamental mathematical parameter of an orbit that indicates the amount of orbital energy contained by an object.

For many centuries, astronomers used the semi-major axis of the Earth's orbit as a useful unit of comparison for objects in the solar system, with the semi-major axis of the Earth defined by $a = 1$ and designated as an *astronomical unit*, abbreviated as AU. Today, 1 AU is defined as the radius of a perfectly circular orbit that an orbiting body would follow if it revolved about the Sun in 365.2568983 days. Using this modern definition, 1 AU = 149,597,870.69 km, which is slightly less than the average distance between the Earth and the Sun.

From the time of Copernicus, astronomers have known the *relative* sizes and distances between the planets in units of AU. Over the centuries, astronomers have found more and more precise ways of determining the absolute value of the AU. The first such measurement was made in 1672 by Jean Richer and Jean-Dominique Cassini, who measured the parallax of Mars and determined that 1 AU was equal to 140 million km. A significant improvement in our knowledge of the length of the AU came through measurements of the solar transits of Venus in 1761 and 1769. By the late nineteenth century, measurements of solar parallax yielded an even more accurate value for the length of an AU, 149.5 million km. Our current knowledge of the length of an AU (149.6 million km) comes from radar echoes off the planet Venus and telemetry from interplanetary space probes.

Johannes Kepler tried to solve the problem of the spacing of the planets using geometric arguments in his first major work, *Mysterium Cosmographicum*. He would go on to establish the correct spacing of the planets in 1619, using his third law. Thereafter, offering explanations as to why the spacing of the planets was that established by Kepler became a cottage industry among German astronomers, as first astronomer Christian Freiherr von Wolff (1679–1754) and philosopher Immanuel Kant (1724–1804) offered their solu-

TABLE 6.1
The Titius-Bode Rule

N	Starting Number	Add 4	Divide by 10	Actual Semi-major Axis	Planet
1	0	4	0.4	0.39 AU	Mercury
2	3	7	0.7	0.72 AU	Venus
3	6	10	1.0	1.00 AU	Earth
4	12	16	1.6	1.52 AU	Mars
5	24	28	2.8	2.80 AU	?
6	48	52	5.2	5.20 AU	Jupiter
7	96	100	10.0	9.56 AU	Saturn
8	192	196	19.6	19.6 AU	?
9	384	388	38.8	38.8 AU	?
10	768	772	77.2	77.2 AU	?

tions. In 1766, Johann Daniel Titius (1729–1796), Professor of Mathematics at Wittenberg, added the following footnote to a book he was editing:[1]

> Given the distance from the Sun to Saturn as 100 units; then Mercury is distant 4 such units from the Sun; Venus, 4 + 3 = 7 of the same; the Earth, 4 + 6 = 10; Mars by 4 + 12 = 16 etc. But see, from Mars to Jupiter there comes forth a departure from this so exact progression. From Mars follows a place of 4 + 24 = 28 such units, where at present neither a chief nor a neighboring planet is to be seen. And shall the Builder have left this place empty? Never![2] Let us assume that this space belongs to the still undiscovered satellites of Mars. . . . Next to this still unexplored space lies Jupiter's sphere of influence at 4 + 48 = 52 parts; and that of Saturn at 4 + 96 = 100.[3]

The position of each planet is determined by doubling the second number, from 3 to 6 to 12 to 24 to 48 to 96, always adding this number to the starting value of 4. If we divide the distance from the Sun to Saturn into one hundred equal parts, with each part equal to 0.1 AU, the planets are found at 0.4, 0.7, 1.0, 1.6, 5.2, and 10.0 AU, with one planet missing at 2.8 AU.

An alternative way to express this idea, and one that can be presented in the form of an equation, is to let N be the planet number, with $N = 1$ for Mercury, $N = 2$ for Venus, with N increasing by one with each planet outward from the Sun. Then we can write that the distance d to planet number N, measured in units of AU, is:

$$d = 0.4 + 0.3 \times 2^{N-2}.$$

The catch, when using this equation, is that it is not used to find the distance for Mercury. Thus, we simply have to assign to Mercury ($N = 1$) the position at $d = 0.4$ AU. We then calculate that for $N = 2$, $d = 0.4 + 0.3 = 0.7$ AU[4], for $N = 3$, $d = 0.4 + 0.3 \times 2 = 1.0$ AU, for $N = 4$, $d = 0.4 + 0.3 \times 4 = 1.6$ AU, with subsequent planets found or predicted to exist at $d = 2.8$ AU for $N = 5$, at 5.2 AU for $N = 6$, and at 10.0 AU for $N = 7$.

For comparison, the actual positions of the planets are 0.39 AU (Mercury), 0.72 AU (Venus), 1.00 AU (Earth), 1.52 AU (Mars), 5.2 AU (Jupiter), and 9.6 AU (Saturn). The matches between the predicted and actual positions are not perfect but certainly are close enough to be intriguing, especially because the predictions can be obtained from an equation, making this relationship akin to Kepler's and Newton's laws.

Six years after Titius published his book with this obscure footnote, Johann Elert Bode (1747–1826) included Titius's idea in the second (1772) edition of his book *Deutliche Anleitung zur Kenntniss des gestirnten Himmels* (Introduction to Knowledge of the Starry Heavens). Whereas Titius is a footnote to astronomical history, Bode was a titan. His *Anleitung* has been described as the most widely read popular work in astronomy in Germany for the better part of a century.[5] In addition, Bode was editor of one of the most important professional astronomy journals, *Astronomisches Jahrbuch*, for over fifty years and became director of the highly respected and influential Berlin Observatory in 1786. Bode's presentation of Titius's idea was almost identical to that of Titius; however, Bode suggested that a heretofore unseen planet, rather than "undiscovered satellites of Mars," ought to exist in an orbit characterized by having a semimajor axis of 2.8 AU.

Is the Titius-Bode rule a mathematical law of physics, on par with Kepler's and Newton's laws, or is it mathematical alchemy? This rule predicts the existence of a planet at 2.8 AU and by extension also predicts the existence of another planet beyond Saturn, at 19.6 AU (for $N = 8$). Does a planet exist between the orbits of Mars and Jupiter? Could a trans-Saturnian planet be found? Few astronomers at the time felt that this mathematical model was sufficiently accurate in predicting the positions of the known planets to justify any claim that trans-Martian and trans-Saturnian planets existed. As a result, though a few astronomers like Bode argued in the 1770s for the existence of a planet at 2.8 AU, at that time no individual or group organized an observational program to search for such a planet and thereby tested the validity of the Titius-Bode rule. The history of science, though, is filled with accidental and serendipitous discoveries and, within a decade of the first appearance of the Titius-Bode rule, one such discovery would usher in two centuries of debate over the Titius-Bode rule and discovery of new objects in our solar system.

THE DISCOVERY OF URANUS

The greatest astronomer of the eighteenth century, Friedrich Wilhelm (William) Herschel (1738–1822), began his adult life by following in his father's footsteps, becoming at age fourteen a musician in a military band in Hanover, Germany. When in 1757 the French army overran his town, he fled to England, where he found employment as an organist and a conductor in a chapel in Bath. This humble church organist also taught and composed music and studied music theory. Music theory has roots in mathematics, and Herschel's talent in music and interest in the mathematical foundations of harmony stimulated his interest in optics and astronomy:

> Among other mathematical subjects, Optics and Astronomy came in turn, and when I read of the many charming discoveries that had been made by means of the telescope, I was so delighted with the subject

that I wished to see the heavens and Planets with my own eyes thro' one of those instruments.[6]

In 1773, Herschel began making his own telescopes and within a few years was able to build telescopes that were bigger and better than those of any professional astronomer of his time. By the end of 1774, he was making regular observations of the heavens.

In 1776, he progressed from a 7-foot-long reflector to a 20-foot-long instrument, the latter with 19-inch-diameter mirrors.[7] Since a large telescope collects more light than a small one, Herschel could see fainter objects in the nighttime sky than other astronomers could. In addition, with the aid of his high-quality optics, the images of the stars in his telescope appeared perfectly round, whereas in other telescopes, with less perfect optics, images of stars appeared distorted.

By 1779, Herschel had taken on one of the toughest tasks in astronomy at the time, the measurement of parallax in order to measure the distances to stars. As we have already discussed, parallax is the apparent change in position of nearby stars, relative to more distant stars, that should occur as a result of the changing viewing position for an earthbound observer, as the orbit of the Earth carries the observer from one side of the Sun to a position 300 million km away on the other side of the Sun. For two thousand years, observers had tried to measure the parallax of stars—and thereby provide experimental proof of the Earth's motion around the Sun—and failed; only in 1838 would Friedrich Wilhelm Bessel (1784–1846) succeed in making the first successful measurement of parallax, of the star 61 Cygni.

The measurement of parallax requires measuring the positions of stars very accurately with respect to the positions of other, nearby stars, but in any field of view as seen through a telescope, only a few bright stars, if any, will appear. Consequently, finding faint stars in the vicinity of bright stars was a key component to making definitive measurements of parallax. Herschel believed, as did many other astronomers at the time, that all stars were identical and that fainter stars were simply more distant objects. Herschel was wrong. Stars

GULIELMUS HERSCHEL LL.D : RSS.

Sidera cuncta notat tacito labentia coelo

Vir. Æneid. Lib.3.

From an Original Picture in the Possession of Wm Watson M.D: F.R.S.

Figure 6.1: Portrait of William Herschel. (By kind permission of the William Herschel Museum, Bath)

intrinsically have different brightnesses, just as lightbulbs have different intrinsic brightnesses. A faint star might be faint because it is far away or because it is simply a dim object. A bright star may appear bright because it is intrinsically luminous or because, despite being intrinsically faint, it is very close. But in the late eighteenth century, the belief that all stars were alike was central to astronomy.

Since a bigger telescope collects more light from a star, just as a bigger bucket of water collects more rain from a storm, more faint stars became detectable in Herschel's large telescope than in smaller telescopes. The task Herschel set for himself was to identify double stars in which one of the two members was much fainter than the other. A double star consists of two stars that appear close together in the sky, though one may be much more distant than the other. If, in fact, stars have intrinsically identical brightnesses, then the faint star in a double star is much farther from Earth than is the bright star. As the Earth orbits the Sun, the position of the more distant star in a double star should not change as much as the position of the closer star; therefore the apparent distance between these two stars should change, so the best double stars to use for parallax measurements would be those with one very bright star and one very faint partner. Thus it was that one evening, during his search for faint stars in the vicinity of one particular bright star in the constellation Gemini, Herschel noticed an object that apparently did not belong in this star field:

> On Tuesday the 13th of March between 10 and 11 in the evening, when I was examining the small stars in the neighborhood of H Geminorum, I perceived one that appeared visibly larger than H Geminorum and the small star in the quartile between Auriga and Gemini, and, finding it so much larger than either of them, suspected it to be a comet.[8]

To Herschel's trained eye, this object was not a star since the image of this object was visibly larger than that of any star. In addition, repeated observations revealed that this newly discovered object was a wandering star, or at least a wandering starlike object, since it was moving with respect to the fixed stars. Since all the planets were

Figure 6.2: Herschel's telescope with which he discovered Uranus. (By kind permission of the William Herschel Museum, Bath)

known already, Herschel assumed the object he had discovered must be a comet; yet its image clearly was round and his observations over months showed no sign of a tail or of any other cometary features.

Herschel reported his unusual discovery to the Astronomer Royal, Nevil Maskelyne, who communicated this news to other observers. Herschel's own paper on his discovery, presented on April 26 to the Royal Society, was entitled "Account of a Comet." Maskelyne, however, was not so sure, suggesting as early as April 4 that this object might be a planet.

Continental observers took up the hunt, with Messier noting that Herschel's object bore no resemblance to any of the eighteen comets he personally had studied. Astronomers at this time expected a comet to have either a parabolic orbit (an "open" orbit such that the object comes from interstellar space, is turned by the Sun, but is not in a permanent orbit around the Sun and so returns to interstellar space) or a highly eccentric orbit (an elliptical orbit with a major axis much bigger than the minor axis). But astronomers at the Observatory of Paris were unable to compute a cometlike orbit that matched the observed positions of the object, and on May 8, Jean Baptiste Gaspard Bouchart de Saron declared that this object was more remote than Saturn, with a minimum distance from the Sun of at least 12 AU. Later that month, Anders Johann Lexell, visiting in London at the time from his home position in Saint Petersburg, calculated a circular orbit for the object with a semi-major axis of 18.93 AU. Although an accurate orbit for the object would require years of additional observations, within three months of Herschel's discovery the scientific world recognized that a new planet had been found. With this discovery, the known size of the solar system had doubled. Given the resolved angular size and the known distance to the object, one can easily estimate the size of the object, and it was immediately evident that Herschel had discovered a fairly large planet as well—a giant planet like Jupiter and Saturn. His own estimate made in 1788 gave Uranus a diameter of just over 55,000 km (only 10 percent larger than the modern value), about 40 percent of the diameter of Jupiter, and 45 percent of the diameter of Saturn but four and a half times larger than that of the Earth.

Without any precedent for the discovery of a new planet, a heated debate, which lasted for over sixty years, began over what to call this object. In deference to his patron, Herschel suggested the name *Georgium Sidus* (George's star) after King George III. This outraged French astronomers who instead called it *Herschel* or *Herschel's planet*. Johann Bode suggested the appellation *Uranus*, which was used by German-speaking astronomers.

So how far from the Sun is Uranus? If we could simply watch the planet for one orbit, determining the distance of a planet from the Sun becomes a straightforward application of Kepler's third law, in which the square of the orbital period is equal to the cube of the semi-major axis. Since an object at a distance of about 19 AU will take about eighty-three years to complete one orbit around the Sun,

Figure 6.3: Hubble Space Telescope view of Uranus at infrared wavelengths. This image reveals the four major rings and ten of the seventeen known satellites of Uranus. (Image courtesy of NASA/STScI)

astronomers would have to wait until the mid-nineteenth century to get an accurate measurement of the size of the orbit of Uranus. Bode, however, realized that the past, as well as the future, could unlock the secrets of Uranus. Rather than wait for observations to be made many decades in the future, he could look into the carefully recorded observing notes of astronomers who might have unknowingly observed Uranus at some time in the past. In this way, Bode was able to uncover prediscovery observations of Uranus made on December 23, 1690, by the first Astronomer Royal John Flamsteed, and on September 25, 1756, by the German astronomer Johann

Tobias Mayer, thereby extending the known observational record for Uranus by ninety years, or more than one full orbital period. In 1787, with these prediscovery observations in hand, Placidus Fixlmiller (1721–1791) calculated that the semi-major axis of Uranus's orbit was 19.18 AU, which is the value accepted today.

AND THEN THERE WERE SEVEN

And so it was that the expert observational astronomer William Herschel discovered, by accident, the first planet unknown to ancient peoples. Now there were seven planets: Mercury, Venus, Earth, Mars, Jupiter, Saturn, and Uranus. Of course, in 1781 the solar system also was known to include ten moons (of Earth, Jupiter, and Saturn) and many, many comets.

With a semi-major axis of 19.18 AU, Uranus is nearly at the position predicted for the $N = 8$ planet by the Titius-Bode rule: 19.6 AU. Does this result for Uranus confirm the Titius-Bode rule as a law of physics? Does the Titius-Bode rule, added to Kepler's three laws, provide a new line of reasoning, based in astrophysics, for determining which objects are planets?

Given Johann Bode's efforts in promoting his rule in the decade preceding the discovery of Uranus, and his impressive detective work in tracking down historical observations in order to determine the orbit for Uranus, he obviously had great interest in finding out whether his rule would become law. The level of accuracy for the position of Uranus (2 percent) is better than the accuracy to which the Titius-Bode rule predicts the positions of Mercury (2.5 percent), Venus (2.8 percent), Mars (5.3 percent), and Saturn (4.6 percent). Astronomers could question the accuracy of the predictive power of the Titius-Bode rule overall, or they could accept that the Titius-Bode rule had in fact predicted the existence of an unknown, unexpected object hidden in the outer reaches of the solar system.

Must we understand why a mathematical formula works in order to apply it? No. Kepler's laws were not understood in 1609 and 1619, when they were first presented; rather, only eighty years later, in

1687 with the work of Isaac Newton, was anyone able to provide a physical basis for Kepler's laws. In 1781, no physical understanding existed to explain why the Titius-Bode rule worked, but apparently it did. And so, in the eyes of many astronomers of the time, the Titius-Bode rule effectively became the Titius-Bode law and entered the pantheon of the great laws of physics. Furthermore, for the first time since Aristotle was dethroned, we now have a definition of a planet—*planets are objects that obey Kepler's laws and the Titius-Bode law*—that clearly distinguishes planets from all other celestial objects, including moons and comets.

The next step is obvious: if the Titius-Bode law successfully predicts the existence of a trans-Saturnian planet, thereby confirming its validity as a law of physics, then the planet predicted by the Titius-Bode law to exist between the orbits of Mars and Jupiter must also exist. Seven planets had been identified, but the Titius-Bode law predicted eight.

The hunt was on.

7
The Celestial Police

O

Perhaps because the Titius-Bode rule was of German origin and therefore a matter of German pride, perhaps because the spacing of the planets was a problem that had been of more interest to German astronomers and philosophers than to those in other countries, and perhaps because more German than non-German astronomers believed that the Titius-Bode rule was physics rather than numerology, the director of the Ernestine Observatory at Seeburg and court astronomer at Gotha, Baron Franz Xavier von Zach (1754–1832), was the one who took up the challenge of finding the missing planet. As his first effort, he tried to compute the position of Bode's object, but the Titius-Bode rule provided information only about distance from the Sun, not about celestial longitude or latitude. Thus, since the missing planet could be in any direction in the sky, von Zach was unable to predict in which direction observers should point their telescopes to search for the predicted eighth planet. Von Zach did make a reasonable assumption in beginning his search: he decided that since the seven known planets had orbits within the zodiac, the missing planet almost certainly would too. Thus, in 1787, working by himself, he began his survey.

Although planets are wanderers among the stars, they move slowly from night to night. How, then, does one recognize which object is the planet, if one is there to be seen? One could watch the same field of stars for an entire night and hope to see one object slowly move; or, one could observe the same field of stars for many consecutive nights and hope to notice one star in a new position with respect to the others; or, one could consult a catalog of all the known stars and simply look for one in the observed field of stars that lacked a counterpart in the catalog, one that was not in its present position when the catalog was made and so must have wandered in from elsewhere. To von Zach, consulting such a catalog seemed to be the best approach, and since no catalog existed, he slowly began to compile one; however, this project was far too hard for a single observer.

Over the next few years, von Zach nearly revolutionized how science could be done. First, in 1796, he convened a conference at Gotha dedicated to discussing methods for finding the missing planet. At this meeting, the French astronomer Joseph-Jérôme de Lalande (1732–1807) suggested that a large number of observers could, together, conduct a systematic, coordinated search for Bode's planet. Such a search would have been an unprecedented, international collaboration to do what, in modern terms, we call "big" science. Nothing came of this idea for several years; however, in late September of 1800, von Zach finally organized this collaborative venture. During a visit to the Observatory of Lilienthal, a private observatory belonging to Johann Hieronymus Schröter, the chief magistrate of Lilienthal, von Zach met with Schröter, Karl Harding, Heinrich Olbers, Freiherr von Ende, and Johann Gildemeister. These six dubbed themselves the Lilienthal Detectives, or the Celestial Police, and dedicated themselves to a joint search for the predicted missing planet.

To this end, they divided the zodiac into twenty-four zones, each fifteen degrees wide. These zones were to be assigned to twenty-four different astronomers—themselves plus another eighteen European colleagues whom they would select and invite into the collaboration. They even elected a collaboration president (Schröter) and secretary (von Zach). One of those to whom they planned to extend an invitation was the Sicilian astronomer Giuseppe Piazzi.

Piazzi ran the Palermo Observatory, the southernmost telescope in Europe. Piazzi did not yet know about the Celestial Police project in late 1800; however, like Herschel before him, he was engaged in a very deliberate and careful project to compile a catalog of stars in order to measure stellar parallax. Ultimately, Piazzi would publish two such catalogs, including one in 1803 with data on 6,748 stars.[1] But first, he would make a spectacular discovery.

CERES DISCOVERED; THE TITIUS-BODE RULE APPARENTLY CONFIRMED AS A LAW OF PHYSICS

On the first of January 1801, Piazzi made his routine nighttime sky observations, measuring the positions of a set of stars in one field in the sky. In order to improve the accuracy of his observations, Piazzi always made several observations of each field he studied. Repeating his work on the second of January, he found that the position he had measured that night for one of his stars was not identical to the position he had measured the previous night. Such a result, he knew, could easily be due to errors in his measurements. By making additional observations on the following night, he could correct what he presumed to be a mistake. But on the third night, he determined, with great certainty, that his measurements were accurate and that the "star" was moving, a wandering star. From three consecutive nights of observations, Piazzi could attempt to predict the direction that the wandering star would travel and estimate how far it would move from night to night. With his predictions in hand, on January 4 Piazzi again easily located his new star and found that it had moved exactly as he had predicted. Clouds—the bane of all earthbound astronomers—prevented him from observing this object again until the tenth, but he was able to find this wanderer on both this night and the following. Amazingly, Piazzi observed the new star on twenty-four nights between January 1 and February 11. In late January, he sent letters announcing his discovery of what he believed to be a comet to the astronomers Lalande and Bode, among others:

On the 1st of January I discovered a comet in Taurus. . . . On the 11th it changed its heretofore (westward) retrograde motion into (eastward) direct motion. . . . I shall continue to observe it and hope to be able to observe throughout the whole of February. It is very small, and equivalent to a star of the eighth magnitude, without any noticeable nebulosity. I beg of you to let me know whether it has already been observed by other astronomers; in this case, I should save myself the trouble of computing its orbit.[2]

Upon receipt of Piazzi's letter, on the twentieth of March, and taking into account the apparent speed of this newfound object through the background of fixed stars, Lalande and Bode immediately embraced Piazzi's discovery as the missing trans-Martian planet.

Proof that this object, which Piazzi tentatively named *Ceres Ferdinandea*, was the sought-for planet would take longer, however. Ceres had by now passed into the daytime sky and, assuming it to be in a near circular orbit close to 2.8 AU, would not be observable again until late summer. To recover Ceres in the autumn skies, astronomers would need to know where to look. By then, Ceres might be far above or below the ecliptic plane, if it were in an orbit inclined, or tilted, to the ecliptic plane. If Ceres were in a highly eccentric orbit, like a comet or even Mercury, its orbital speed might speed up or slow down by a large amount, even in a few months' time. Ceres would be lost unless someone could determine an accurate orbit that would enable future positions of Ceres to be calculated, especially since the information recorded by Piazzi for the positions of Ceres at different moments during January and February was insufficient, given existing mathematical techniques. Indeed, by August observers began looking for Ceres at positions predicted from such calculations, but the predictions were wrong. Ceres was lost.

Fortunately, the discovery of Ceres stimulated the interest of one of the great mathematicians of the nineteenth century, Karl Friedrich Gauss (1777–1855). Gauss learned about Piazzi's discovery in September and, within two months, had developed a new technique for calculating orbits and sent his results to von Zach. On January 1, 1802, exactly one year after the original discovery by Piazzi, von Zach and, independently, Olbers recovered Ceres using the predictions made by Gauss.

The big question now concerned whether Ceres was the missing planet predicted by the Titius-Bode rule. Gauss assigned Ceres a semi-major axis of 2.767 AU. The Titius-Bode rule predicted 2.8

Figure 7.1: Asteroid trail of the very fast-moving object 1994 XM1, obtained in this 150-second time exposure by Jim Scotti of the Spacewatch project on December 9, 1994. The asteroid was 550,000 km from Earth at this time. In modern time-exposure photographs, the motion of asteroids projected against the very distant "fixed" stars produces streaks in such images. Pre–photographic era astronomers like Piazzi had to recognize the motion of a solar system body as it moved to a new position among the stars, when observed on more than one night. (Observation by the Spacewatch Project of the Lunar and Planetary Laboratory, University of Arizona. Copyright 1994 by the Arizona Board of Regents. Reproduced by permission)

AU, so Ceres was right where the Titius-Bode rule predicted it to be; in fact, Ceres was closer to the predicted position than were any of the known planets except Earth and Jupiter! The Titius-Bode rule was confirmed again, but for one small problem. Ceres was small and faint.

Mars and Jupiter, the former of which is closer to Earth than Ceres, the latter of which is more distant from Earth than Ceres, are both among the brightest objects in the entire sky. If Ceres is a planet and is in between Mars and Jupiter, shouldn't it also be bright? If it is not bright, then perhaps it is not big and, by implication, not

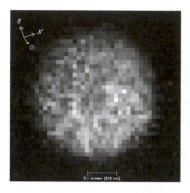

Figure 7.2: The figure on the left is an ultraviolet light image of Ceres, obtained by Alan Stern using the Hubble Space Telescope. The sunlight direction is toward the lower left, as is indicated by the arrows in the top left corner of the image. The figure on the right is a computer-generated model image of Ceres, based on this ultraviolet image. The dark spot near the center is known as the Piazzi feature. (Images courtesy of the Southwest Research Institute and NASA)

planetlike. William Herschel measured the angular size of Ceres and found that it had a diameter of only 258 km. Herschel was wrong about the diameter of Ceres, as Ceres, with a diameter of about 920 km, is almost four times bigger in diameter than this measured value. But Herschel had discovered Uranus and was perhaps the most widely respected astronomer of his time; consequently, his measurement carried great weight within the astronomy community. Given that Mercury, the previous record holder as the smallest planet with a diameter of 4,878 km, is twenty times bigger in diameter and therefore four hundred times larger in cross section and eight thousand times larger in volume than Ceres, assuming the size Herschel determined for Ceres is correct, Ceres hardly looked like a planet, or at least like all the previously known planets.

THE SOLAR SYSTEM GETS CROWDED

The debate over the significance of Ceres heated up when Olbers announced, on March 28, 1802, that he had discovered an object similar to Ceres that he named Pallas. Pallas had a semi-major axis of 2.67 AU and, based on Herschel's measurements, a diameter of only 176 km (the modern value is about 580 km). Could two planets be so close together? Even more disconcerting for the advocates of planet status for Pallas and Ceres were the parameters of their orbits. Ceres has a nearly circular orbit, but its orbit is inclined to the ecliptic plane by more than 10 degrees. Mercury's orbit is inclined by 7 degrees, Venus's by 3.4 degrees, the Earth by definition orbits in the ecliptic plane, Mars's orbit is inclined 1.85 degrees, Jupiter's orbit 1.3 degrees, Saturn's orbit 2.5 degrees, and newly discovered Uranus's orbit only 0.8 degrees. Is an inclined orbit such as that of Ceres planetlike? However problematic the planetary qualifications of Ceres may have been, Pallas was an even poorer specimen. Not only is Pallas's orbit inclined to the ecliptic by almost 35 degrees, it was immediately evident that it had an extremely non-circular orbit, with an eccentricity of almost 0.25.[3] With such a large eccentricity, Pallas would orbit as close to the Sun as 2.0 AU and as far from the Sun as 3.3 AU. In fact, Pallas would cross the orbit of Ceres.

Bode first derided Pallas as a comet and championed Ceres as the sought-after trans-Martian planet that confirmed his prediction. Later, he conceded that both were planets and expressed the opinion that two planets confirmed the validity of his rule just as well as one. In early 1802, did we know of eight planets? Or nine planets? Or only seven planets but two objects of some different type? Herschel thought there might be many more small objects in the large spaces between the planets, and in 1802 he suggested to the Royal Society the name *asteroids* (starlike) for such objects. For better or worse, inappropriate as the name may be, the appellation *asteroids* would eventually stick, but not for more than a century. In the minds of some, by June 1802, less than eighteen months after Ceres had been discovered and only six months after its existence had been confirmed, its orbit determined, and its status as Bode's missing planet proclaimed, this new object had been demoted to something less than a planet, and an entirely new class of solar system objects had been revealed. According to this point of view, the solar system now had planets, moons (or satellites), comets, and asteroids.

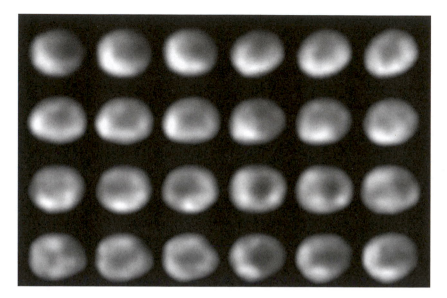

Figure 7.3: Series of Hubble Space Telescope images of Vesta, showing Vesta at twenty-four different positions during the full 5.34-hour rotation of the 525-km-diameter asteroid. The smallest resolvable features in these images are 80 km across. Features include ancient lava flows and a gigantic impact basin that is so deep that it exposes the asteroid's subsurface, or mantle. Images were obtained with the Wide Field Planetary Camera 2 between November 28 and December 1, 1994, when Vesta was at a distance of 156 million miles from Earth. (Image courtesy of B. Zellner, Georgia Southern University, and NASA)

This understanding of Ceres and Pallas was not accepted by most astronomers, however. To most, the solar system now had nine planets. Soon, it would have ten, then eleven.

Barely two years after the discovery of Pallas, on September 1, 1804, another of the Celestial Police, Harding, discovered a third object in the vicinity of the first two, Juno (diameter 249 km, semi-major axis 2.67 AU), and on March 29, 1807, Olbers discovered a fourth, Vesta (diameter 555 km, semi-major axis 2.36 AU). If basic astronomy textbooks of the first half of the nineteenth century are a measure of consensus, the solar system indeed had eleven planets for several decades. Even William Herschel's son John, himself one of the most prominent astronomers in the world in the early nineteenth century, counted eleven planets, with Vesta, Juno, Ceres, and

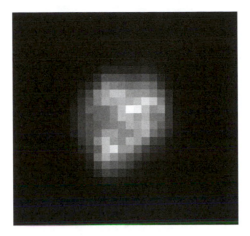

Figure 7.4: Image of Juno in blue light taken in 2003 with the one-hundred-inch Hooker Telescope at Mount Wilson Observatory in four different colors. This image shows that Juno is misshapen rather than spherical. (Image courtesy of Sallie Baliunas and the Harvard-Smithsonian Center for Astrophysics)

Pallas counted as the fifth through eighth planets, and Jupiter renumbered as the ninth, Saturn the tenth, and Uranus the eleventh.

Four decades would pass before, on December 8, 1845, the amateur astronomer Karl Ludwig Hencke would add to the list a twelfth planet (now considered the fifth asteroid), Astraea (2.57 AU, 120 km in diameter). For a brief time, Ceres, Pallas, Juno, Vesta, and finally Astraea enjoyed the status of being the eighth, ninth, tenth, eleventh, and twelfth known planets. But these objects were too small, their orbits too unusual, and soon they would have too much company. Of greatest significance, a thirteenth planet, Neptune, this one a large planet, would be discovered in 1846.

By 1847, greatly influenced by the 1846 discovery of Neptune, William Herschel's designation of asteroids had gained favor; at best, Ceres, Pallas, Juno, Vesta, and Astraea were *minor* planets, asteroids or planetoids. The astronomy community would soon revoke the label "planet" from these five objects. From seven to eleven to twelve to thirteen and back again to only eight planets, the first half of the nineteenth century was a tumultuous time for planets.

As for the asteroids themselves, the dam broke in 1847. With the more powerful telescopes and better star charts then available,

three more asteroids were discovered that year, followed by one discovery in each of the next two years, making a total of six asteroids
discovered in the 1840s. Subsequent decades saw 47 discovered in
the 1850s, 52 in the 1860s, 102 in the 1870s, 76 in the 1880s, and
165 in the 1890s. Within a century of the discovery of Ceres, 463
asteroids had been discovered and their orbits determined. Now,
after two centuries, several hundred thousand asteroids have known
orbits, and tens of thousands more are discovered and added to this
count every year.

But what of the Titius-Bode rule? Did the discovery of Ceres
confirm or did the discoveries of Pallas and Juno invalidate the
Titius-Bode rule as a law of physics? By 1805, the Titius-Bode rule
apparently had successfully predicted the discovery of two (Uranus
and Ceres), perhaps even four (Pallas, Juno), objects. Uranus was
about where the Titius-Bode rule predicted it to be, and Ceres, Pallas, and Juno were fairly close to the predicted position for a trans-
Martian planet. Olbers made the suggestion that perhaps a trans-
Martian planet had once existed, thereby robustly confirming the
Titius-Bode rule, but that a later event, perhaps one or several collisions or explosions, destroyed the planet. Thus, the asteroids, he
suggested, might be fragments of the original large planet. Late nineteenth-century calculations by the French mathematician Urbain
Jean Joseph Le Verrier suggested the aggregate mass of the whole
swarm of asteroids could not exceed one-fourth of the mass of the
Earth; such a planet, though small, would have been more massive
than both Mercury and Mars put together. Others, however, estimated that much less mass, perhaps less than 1 percent of the mass
of the Earth, was present in the zone of the asteroids. Such a small
mass was hardly enough to reconcile with the idea that the asteroids
were fragments of an exploded planet.

The second explanation offered for the origin of the asteroids,
which is the explanation that has become the modern consensus, is
that no single large object ever existed in the region we now call the
asteroid belt; the largest asteroids are thought to be leftover building
blocks of planets—planetesimals—rather than fragments of a fractured planet. According to this explanation, due to the perturbative,

or disturbing, gravitational influence of Jupiter, no single object in the region of the solar system between Mars and Jupiter ever grew big enough to become as large as Mercury or Mars. The powerful attraction of Jupiter pulled on objects in the asteroidal region of the disk around the young Sun and stirred up their orbits, making their orbits more eccentric and highly inclined to the ecliptic plane. Consequently, collisions between asteroids during the planet-formation epoch tended to occur at higher speeds than collisions in other parts of the solar system, resulting in destructive collisions rather than growth of ever larger objects. Nevertheless, in the early nineteenth century, not only was Olbers's suggestion a reasonable bit of speculation, it gave astronomers who thought that the Titius-Bode rule had been confirmed as a law of physics an additional reason to continue to give it the benefit of the doubt.

If the Titius-Bode rule successfully predicted the existence of a trans-Martian planet (albeit a fractured one) at 2.8 AU and a trans-Saturnian planet at 19.6 AU (albeit at 19.18 AU), what else could the Titius-Bode rule predict? For $N = 9$, it predicts the existence of a planet at 38.8 AU.

Could early nineteenth-century astronomers find such a planet? The Celestial Police felt they needed twenty-four astronomers to commit to a dedicated, collaborative search for a planet at 2.8 AU. How many would they need in order to organize a search for a hypothesized planet 14 times farther from the Sun? An object 38.8 AU from the Sun would receive 256 times less light than if it were only 2.8 AU distant. In order for us to observe this object in a telescope, the light from this same object would have to reflect back from 38.8 AU to the Earth, making it an additional 400 times fainter than the reflected light from an identical object located at 2.8 AU. Thus, simply based on the inverse square law for light, the Celestial Police would be looking for an object located 38.8 AU from the Sun that would appear nearly 100,000 times fainter than an object only 2.8 AU from the Sun, assuming the objects at 38.8 AU would be similar in size to Ceres and Pallas. The discovery of such faint objects was not even conceivable in the early nineteenth century; however, if the Celestial Police were searching for a single large object like Ura-

CHAPTER 7

nus, but at 38.8 AU rather than at 19 AU, that large, distant object would be only 16 times fainter than Uranus. In addition, knowing that Ceres and Pallas had inclinations to the ecliptic greater than those expected, while Uranus was nearly in the ecliptic, such a planet-hunting team would have to consider searching parts of the sky farther above and farther below the ecliptic than was anticipated in the plans for the first trans-Martian planet search. Certainly, far more than twenty-four astronomers, all of them using bigger tele-scopes than they currently had, would be a big international collabo-ration and would require a substantial investment in new equipment and possibly even the development of new observatories dedicated to the project; such a project was not even remotely possible in 1810. Thus, if a trans-Uranian planet were to be discovered, it would be through serendipity, unless a new method could be developed that would focus astronomers' attention on only a small portion of the night sky.

106

8
Neptune, the Thirteenth Planet

O

The discovery of Uranus ushered in an era of excitement without precedent in astronomy. Even if the Ptolemaic and Aristotelian models of the universe, with their perfectly circular, geocentric orbits, were wrong, the orbits of the Sun, Moon, Mercury, Venus, Mars, Jupiter, and Saturn had been fairly well-known since the times of Hipparcus and Ptolemy, 1,800 years earlier, at least in the sense that astronomers could use mathematical models to predict future positions of those objects. Now, at last, astronomers had a new object to study, a new orbit to map out, and they took on this challenge with great gusto.

As we have already seen, Bode invested a great deal of effort to uncover two prediscovery observations of Uranus, one made in 1690 by Flamsteed, the second in 1756 by Mayer. When the 1690 and 1756 observations were combined with those of Herschel and others from the 1780s, the orbit of Uranus could be calculated to a reasonable accuracy. Yet, by 1788 Uranus was straying surprisingly far from the positions predicted for it. In 1788, Pierre Charles Lemonnier discovered that he had made three prediscovery observations, in 1764 and 1769.[1] Perhaps these additional data points would fix the apparent errors in the predictions for Uranus's orbit. But even with a total of five prediscovery observations in hand, the astronomers of the time found that Uranus's motion defied their best efforts to calculate a single orbit that fit both the prediscovery and postdiscovery observations. By early in the nineteenth century, it was clear that

107

the historical and new observations of Uranus could not be reconciled with the orbit that astronomers predicted for it, assuming Uranus follows an elliptical orbit around the Sun while that orbit is constantly being tweaked by the gravitational tugs of the known planets out to Saturn. The errors in the observed positions of Uranus were never off by more than two minutes of arc from the predicted positions, but even this small error was fully ten times larger than observational errors normally attributed to Flamsteed's seventeenth-century observations, let alone the more accurate eighteenth- and early nineteenth-century observations.

By 1820, a total of seventeen prediscovery observations of Uranus had been uncovered, but these data only exacerbated the problem of deriving a single, sensible orbit for Uranus. What was wrong? In 1821, Alexis Bouvard took the approach of asserting that the ancient observations were wrong and, thus, should be discarded as bad data, although he added the prescient proviso that future investigators may determine that the inaccuracies of the ancient observations depend "on some extraneous and unknown influence which may have acted on the planet."[2]

Bouvard's decision to discard data that did not easily fit the 1771–1820 epoch orbit he deduced for Uranus is not a respectable way to carry out scientific research. We do discard data if we can demonstrate that those measurements are flawed, but we do not discard measurements simply because the data do not fit our models; instead, we recognize that we must adjust our models and our understanding to match the data. Even worse for Bouvard, those of us with twenty-twenty hindsight now know that the prediscovery observations were quite accurate. In fact, Bouvard had to assign errors of over a minute of arc (more than sixty seconds of arc) to observations recorded by distinguished astronomers whose observations were generally assumed to be ten times more accurate (with errors of only six to ten seconds of arc). Bouvard, however, was in a quandary. In order to predict where Uranus might be in future years, he could not use the prediscovery data, no matter the reason for discarding it.

Yet, by 1825 even Bouvard's work using only the postdiscovery observations failed. Uranus defied all efforts to predict its behavior. Then, for a period of a few years, Uranus began to behave better again. Maybe, after all, the calculations were right, the observations were right, and Newton's law of gravity was right.

Then, in the late 1820s the observed positions of Uranus began to pull away again from the predicted positions. By 1832, Uranus was the width of the full Moon (thirty minutes of arc) away from the position predicted for it. The astronomers of this era might not have been perfect, but they did not make errors in measuring positions of celestial objects by anything close to this angular size. Something was clearly wrong with Uranus or with the law of gravity.

To understand what happened over the next decade, we must step away from this specific problem and review in more detail the issues an astronomer must take into account when calculating the orbit of a planet like Uranus. The simplest possible problem when dealing with orbits and gravity is known as the two-body problem. In this simple, hypothetical situation, only two objects—for example, the Sun and the Earth—exist. Then the problem of calculating the orbit of the Earth around the Sun is very simple and can be solved exactly; however, the solar system has more than two objects in it.

What happens if we add Jupiter to our simplified solar system? The mass of Jupiter is one thousand times smaller than the mass of the Sun, so, all other things—like the distances from the Earth to the Sun and Jupiter—being equal, the gravitational influence of Jupiter on the Earth would be one thousand times smaller than the gravitational influence of the Sun on the Earth. But all other things are not equal. When Jupiter and the Earth are on opposite sides of the Sun, Jupiter is six times farther from the Earth (1 AU + 5 AU = 6 AU) than is the Sun (1 AU). Since the gravitational force is inversely proportional to the square of the distance between two objects, at such a time the gravitational influence of Jupiter on the Earth is thirty-six times smaller than that of the Sun, all other things (i.e., the mass of the two objects) being equal. When Jupiter and the Earth are on opposite sides of the Sun, the greater distance from

Earth to Jupiter than from Earth to the Sun, combined with the mass of Jupiter being so much smaller than the mass of the Sun, results in the gravitational influence of Jupiter on the Earth being thirty-six thousand times weaker than that of the Sun on the Earth. When Jupiter and the Earth are on the same side of the Sun, Jupiter can be as close to the Earth as about 4 AU, only four times farther than the Sun. In this three-body configuration, Jupiter's gravitational influence on the Earth is only sixteen thousand times weaker than that of the Sun. While Jupiter's tugs on the Earth are significantly weaker than those of the Sun itself, these tugs are measurable and significant in affecting the orbit of the Earth over large periods of time. In fact, long-term, oscillatory changes in the shape of the orbit of the Earth caused by Jupiter are the primary causes of terrestrial ice ages.

In the real solar system, even the three-body problem is too simple: we must also include the gravitational effect of Venus on the Earth. Venus is almost 400 times less massive than Jupiter, yet Venus can come as close as 0.27 AU from Earth. The small mass of Venus reduces its gravitational influence on the Earth compared to the gravitational influence of Jupiter, but by virtue of getting 15 times closer to Earth than the closest approach of Jupiter, Venus partially compensates for its low mass with an advantage of a factor of 225 over Jupiter. Thus, when Venus is closest to Earth, its gravitational influence is about 56 percent (the combined effect of being 225 times stronger but also 400 times weaker) times that of Jupiter, when Jupiter is exerting its most powerful influence on Earth. Clearly, if the gravitational tugs of Jupiter cannot be ignored, the influence of Venus also must be taken into consideration. Now, any calculation of the future position of the Earth has become a four-body problem. To perform the necessary mathematical calculations, one must know the positions of Venus, Jupiter, and the Earth at all times, since the relative influences of Venus and Jupiter on the Earth change as a function of time, as the three planets orbit the Sun at different speeds.

To calculate the orbit of Uranus, astronomers like Bouvard had to account for the perturbing gravitational influence on Uranus of all the known planets, especially those of Jupiter and Saturn. This calculation was a very difficult problem, but nevertheless one that astronomers in the first half of the nineteenth century were capable of solving very accurately, assuming Newton's law of gravity was correct. Given that Uranus continued to follow an orbit that confounded the best orbital calculations, either something was wrong with the law of gravity or some unseen, unknown object was exerting a gravitational pull on Uranus, constantly perturbing Uranus out of the predicted orbit. This was the meaning of the offhand suggestion made by Bouvard in 1821. By 1834, others had taken up this idea.

Amazingly, the slightly delayed return of Halley's comet in 1835 led to a similar suggestion. Comet Halley reached perihelion, the closest point in its orbit to the Sun, one day later than expected, according to the calculations of Friedrich Bernhard Gottfried Nicolai (1793–1846), director of the Mannheim Observatory. The predicted date of perihelion was based on a many-body calculation of the orbit of Comet Halley, taking into account the gravitational influences of all the known planets; consequently, some force, presumably the gravitational influence of an unknown object in the outer solar system, must have tweaked Halley's orbit just a little bit over the previous seventy-six years. Both Jean Eliz Benjamin Valz (1787–1861) at Marseilles, who made similar calculations and predictions, and Nicolai suggested that a trans-Uranian planet, probably at a distance of about 38 AU as predicted by the Titius-Bode rule, was likely the cause of the unexpected change in the orbit of Comet Halley.[3]

By the early 1840s, the problem of the unexplained discrepancies in the orbit of Uranus had become one of the most important in the physical sciences. Either Newton's law of gravity was right, in which case the problems in predicting the orbit of Uranus could now be used to predict and discover another new planet, or Newton's law of gravity was wrong, and these same problems could lead to the discovery of a better, more accurate law of gravity. Whatever the explanation, the outcome would be both momentous and financially rewarding, as in 1842 the Royal Academy of Sciences at

Gottingen offered a prize to the scientist who provided the solution to the Uranus problem.

In 1842 Friedrich Wilhelm Bessel, who in 1838 became the first astronomer to successfully measure the parallax of a star, thus achieving the holy grail accomplishment sought after by astronomers for over two thousand years, took on the challenge of the problematic orbit of Uranus, declaring that the prediscovery observations of Uranus were likely very accurate and could and should be used to predict and "discover the orbit and mass for some unknown planet."[4] Bessel did not get far in solving the Uranian problem, as his student, who was to do the calculations, died and then he himself became ill and remained in poor health until his death in 1846. However, two other mathematicians took up the challenge, John Couch Adams of England and Urbain Le Verrier of France.

Arguing in generalities that the differences between the predicted and observed orbital positions of Uranus were due to the perturbing influence of a more distant, heretofore unknown planet is easy; however, going from the general assertion to a specific answer and prediction for the size, mass, and location of the mystery planet is an extremely formidable challenge. In order to solve the problem, one would have to guess a longitude around the Sun for the unknown Planet X at a certain date and, further, assign a mass to Planet X and orbital parameters (semi-major axis, eccentricity, and inclination) to constrain the orbit of Planet X.[5] The orbital parameters determine, for all past and future times, the position of Planet X. The position of Planet X relative to Uranus at any moment in time, combined with the mass of Planet X, determines the gravitational influence, both in strength and direction, of the extra perturbing force on Uranus by Planet X.

This type of mathematical challenge is known as an inverse problem: we know the effect, and from that effect we must ascertain the cause. Predicting an effect from a known cause is much easier than determining an unknown cause from a measured effect. A century before the invention of the computer, this was an enormous challenge solvable only with keen insight to reduce the infinite number of possible solutions to the few worth calculating.

In the period from 1843 to 1845, Adams analyzed all the existing orbital data for Uranus, searching for a unique solution. By the middle of 1845, having given up on all possible circular orbits for Planet X, he found an answer. On the twenty-first of October 1845, he sent the details of his solution to the Astronomer Royal, George Biddell Airy. Inspired by the Titius-Bode rule, which predicted a planet at 38.8 AU, Adams predicted that the trans-Uranian planet would have a semi-major axis of 38.4 AU, an eccentricity of 0.16, and a mass of 49.7 times the mass of the Earth. He also provided a heliocentric longitude for October 1, 1845, thus providing a complete set of orbital parameters that would enable an observer to search for and presumably find this planet. All that was needed was for someone to point a telescope in the predicted direction and look. Yet although Adams had just been elected to membership in the Royal Astronomical Society and had extremely impressive credentials, having won every possible award available to him as a student at Cambridge in the early 1840s, the Astronomer Royal apparently did not take the youngster's prediction seriously and did not push his staff to make the necessary observations. Adams would continue to refine his calculations over the next year, but for nearly a year, Airy refused to initiate a search for Planet X.

As it turned out, Adams had competition on the continent. In 1846, Le Verrier independently made a similar mathematical prediction as Adams. In the last week of June 1846, eight months after Adams had offered Airy his mathematical prediction for finding Planet X, Airy exchanged letters with Le Verrier, in which Le Verrier explained his methods and provided part, but not all, of his solution. The similarity of Le Verrier's and Adams's solutions finally spurred Airy into action. On July 9, using his authority as Astronomer Royal, he instructed James Challis, the Plumian Professor of Astronomy at Cambridge and the director of the Cambridge Observatory, to begin a search for the hypothesized planet. Challis's search, however, was plagued by cloudy weather and a conservative search technique. Rather than immediately look at the position in the sky predicted by Adams, Airy instructed Challis to survey a 30 degree by 10 degree area of the sky, centered on the predicted position, map-

Figure 8.1: *Voyager 2* spacecraft image of Neptune, showing white high-altitude clouds. (Image courtesy of NASA/JPL-Caltech)

ping all the stars in each nine-arc-minute field of view. By Challis's reckoning, the procedure demanded by Airy would take Challis thousands of hours of observing time to complete. Challis made his first observations on July 29.

Le Verrier finally made public his exact solution at the end of August: semi-major axis of 36.154 AU, an eccentricity of 0.17, and a mass of 32.3 times the mass of the Earth. Le Verrier made very clear in his predictions exactly where to look to find his planet and what his planet would look like. No French astronomers took up the challenge. No one looked.

Finally, Le Verrier contacted a young colleague at the Berlin Observatory, Johann Gottfried Galle. Galle received from Le Verrier, on September 23, a prediction for the position of the never-before-seen planet. After receiving permission (against his better judgment) from the observatory director, Johann Encke, Galle, with the observing assistance of Heinrich Louis d'Arrest, and using star map XXI in the new stellar atlas of Carl Bremiker,[6] looked for and found Neptune that very night. Less than three weeks later, on

October 10, the Liverpool beer-brewer-turned-astronomer William Lassell (1789–1880) discovered Neptune's largest moon, Triton.

Neptune was nearly the exact size, in appearance, as predicted by Le Verrier and was less than one degree—fifty-five minutes of arc—from the position predicted for it by Le Verrier. As it turns out, Adams's predicted position was almost as good, off by less than one and one-half degrees (or ninety minutes of arc). Ironically, despite the ridiculous search technique demanded by Airy, Challis had observed Neptune on August 4 and 12, but had not analyzed his sketches carefully enough to recognize that one of his "stars" had moved.

The discovery of Neptune ushered in a number of important changes in solar system astronomy. With the discovery of Neptune, the solar system had grown enormously, again. Prior to the discovery of Uranus, the entire solar system was contained within 10 AU of the Sun. The discovery of Uranus doubled the size of the solar system from 10 to 20 AU. The discovery of Neptune added yet another 10 AU to the radial size of the known planetary system.

Was Neptune the thirteenth planet known in the solar system? Amazingly, the answer is both yes and no. By the time of the discovery of Neptune, in late 1846, astronomers had had forty years to digest the earlier discoveries of Vesta, Juno, Ceres, and Pallas, the alleged eighth, ninth, tenth, and eleventh planets in the solar system. Then, less than a year before the discovery of Neptune, a fifth trans-Martian object and the twelfth planet, Astraea, had been found. Thus, when Neptune was discovered in late 1846, it took its place in line as the thirteenth planet. Yet, by 1847 William Herschel's decades-old suggestion that the trans-Martian objects are minor planets, or asteroids, and not major planets, took hold. In the minds of many, 1847 was a watershed year during which Ceres, Pallas, Juno, Vesta, and Astraea were reclassified as asteroids. Neptune moved up in line: it was both the thirteenth planet, including major and minor planets, and the eighth major planet.

With the discovery of Neptune, the solar system now had four giant planets in the outer solar system to go along with the four smaller planets in the inner solar system and at least five asteroids

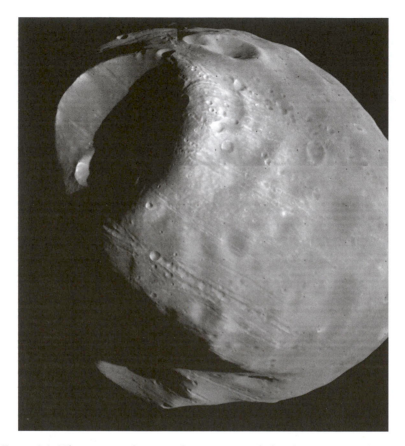

Figure 8.2: *Viking* spacecraft image of Mars's moon Phobos (meaning "fear"). (Image courtesy of NASA and the NSSDC)

in between. By 1849, the solar system was populated by eight planets, ten asteroids, fourteen moons,[7] and uncounted numbers of comets. Many more asteroids would be discovered in the coming decades, with nearly five hundred known by the end of the nineteenth century. American astronomer Asaph Hall discovered Mars's two small moons, Phobos and Deimos, in 1877, leaving only Mercury and Venus, among the eight planets, lacking moons.

The most immediate change in the physics of the solar system involved the Titius-Bode rule. Neptune did not have an orbital semi-major axis of 38.8 AU, as predicted by Bode's rule, nor was it as far away as 38.4 AU, as predicted by Adams, or even 36.15 AU

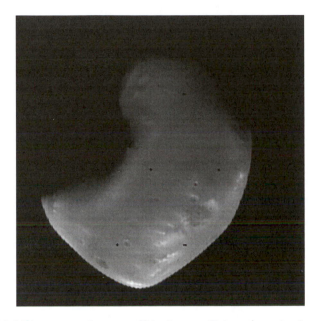

Figure 8.3: *Viking* spacecraft image of Mars's moon Deimos (meaning "panic"). (Image courtesy of NASA and the NSSDC)

as predicted by Le Verrier. By October 1846, Adams had calculated Neptune's orbit from the observational data and found that the semi-major axis was only 30.05 AU. Galle's colleagues at the Berlin Observatory reported 30.04 AU. With preliminary orbits in hand, astronomers commenced their searches for prediscovery observations and were rewarded with two observations recorded by Lalande from May 8 and 10, 1795, with which the American astronomer Sears Cook Walker calculated a more refined orbit for Neptune: semi-major axis 30.25 AU, mass 20 Earth masses. As it turns out, the earlier orbit calculations were closer to the mark: the semi-major axis of Neptune is 30.06 AU, but Walker's mass calculation is closer to the modern value of 17 Earth masses.

The Titius-Bode rule erred in predicting the distances of Mars and Saturn by about 5 percent and of Mercury, Venus, and Uranus by about 2 percent. Furthermore, it predicted one planet, not a multitude of tiny objects, between the orbits of Mars and Jupiter. All these errors had been dismissed as small, as inconsequential, permitting believers to ascribe to the Titius-Bode rule powers it did not merit. Now, however, the Titius-Bode rule was off by more than 8 AU, almost 28 percent. This error was much too large to be acceptable. The astronomers of the mid-nineteenth century who were supporters of the Titius-Bode rule were forced to accept that this mathematical formula had failed its most important test: it could not be used to make accurate predictions, one of the hallmarks of a good scientific theory. The Titius-Bode rule was not a law of physics that provided insight into the spacing of the planets and made predictions for the existence of yet undetected planets; it was merely a neat mathematical trick, now to be abandoned to the rubbish heap of history.

Astronomers in the mid-nineteenth century were not concerned with the fact that, without the Titius-Bode rule, they had lost any physically based rationale for understanding what was or was not a planet, but they had. Without the Titius-Bode rule as a law defining planets as objects that satisfy Kepler's laws *and* fall at predetermined orbital distances from the Sun, the only way to determine that Ceres and Comet Halley were an asteroid and a comet, respectively, and not planets was consensus. The problem with consensus is that it eventually breaks down.

Aristotle had his wandering stars made of aether; such objects uniquely defined planets in an Aristotelian universe. But the Aristotelian universe collapsed with the arrival of the Copernicans.

Early Copernicans, prior to the discovery of asteroids and comets, could uniquely define planets as those objects in heliocentric orbits.

By 1620, with asteroids unknown and comets still not understood as objects in orbit around the Sun, astronomers could define planets as the only objects that satisfied Kepler's laws. This status quo held firm for well over a century.

But Halley's comet, being in a heliocentric orbit, complicated the situation: planets were no longer the only heliocentric bodies. Comets were not an enormous problem, however, because such objects had been known as comets for thousands of years. Everyone knew they were not planets. Planets could be defined as all the objects in heliocentric orbits that were not comets.

Then, in the late eighteenth century, the discovery of Uranus and the sanctification of the Titius-Bode rule as law provided a means to uniquely define planets and distinguish them from comets. While the discovery of the asteroids complicated matters, astronomers could brush this problem under the rug by suggesting that asteroids were planets or fragments of a broken planet.

But now, with the Titius-Bode rule dismissed as numerology, for the first time since Aristotle astronomers were left without any firm basis in physics for defining a category of objects called planets. Perhaps planets are like beautiful sunsets: we all know one when we see one. Implicitly using this aesthetic, nonphysical criterion, everyone was satisfied. Neptune obviously was a planet, as were Mercury, Venus, Earth, Mars, Jupiter, Saturn, and Uranus. Asteroids and comets and moons obviously were not planets. What more do we possibly need to know?

On the positive side, Newton's law of gravity had been shown to be enormously powerful and successful, as it had been used to calculate the position of Neptune based on Neptune's gravitational influence on Uranus. The fact that the real Neptune was closer and less massive than the Planet X calculations of either Adams or Le Verrier demonstrates the difficulty of solving an inverse problem: often, multiple solutions exist. For example, $2 \times 2 = 4$. Thus, four is the unique solution to the question "What is two times two?" However, $-2 \times -2 = 4$ also. Thus, if we try to solve the inverse problem by asking "What is the square root of four?" the answer is "Either $+2$ or -2." The solutions of both Adams and Le Verrier placed Neptune in the right direction but at the wrong distance from the Sun; however, a less massive planet placed closer to the Sun but in the same direction as a more massive, more distant planet will have a similar gravitational effect on Uranus, in both strength and direction, at least for a period of a few decades, after which the different orbital periods of closer and more distant planets would begin to affect the calculations.

Now that the problems with the orbit of Uranus had been solved, the next phase in solar system discovery was obvious: since the orbit of Uranus had been used to predict the existence and ultimately the location of Neptune, perhaps the orbit of Neptune could be used to find the next Planet X. If this harder inverse problem could be solved—harder because accurate measurements of the position of the more distant Neptune are harder to make than those for closer planets and because the location of an even more distant planet is even harder to constrain than was the position of Neptune, especially without help from the Titius-Bode rule—then the expected trans-Neptunian planet might be found.

9
Easy Come, Easy Go

O

Although the discovery of Neptune revealed one law of physics to be an impostor, it cemented the status of Newton's law of gravity. In addition, the method by which Neptune was discovered clearly suggested that the same method could be used to predict the existence and location of the next planet beyond Neptune. Astronomers merely needed to observe the location of Neptune over a period of years, note the discrepancies between its observed and predicted positions, and use those discrepancies along with the new mathematical tools developed by Le Verrier and Adams to determine the mass and location of the ninth planet, the new Planet X.

In 1848, Jacques Babinet made the first such prediction of a trans-Neptunian planet, one with a mass 11.6 times that of the Earth at a distance of 47–48 AU. Neptune, which has an orbital period of 165 years, had moved only about 4 degrees in the sky, out of a single 360-degree orbit, which was hardly enough motion to merit a declaration that the "known" orbit was flawed and that a Planet X was the cause. While Neptune's position was known back to at least 1795, using prediscovery observations, Babinet's prediction was based only on the differences between the orbit predicted for Neptune by Le Verrier and that for the actual planet. Babinet's prediction, which was based more on what was expected rather than what the data indicated, was generally ignored. Nevertheless, his prediction is a measure of the intellectual climate of the time; he was merely the first to try his hand at using Neptune to predict the location of Planet X.

The first reported discovery of a trans-Neptunian planet was made in 1851, by Englishman John Russell Hind, though Hind himself credited the discovery to a staff astronomer at the U.S. Naval Observatory, James Ferguson. Hind had read a report, published by Ferguson in the *Astronomical Journal* in January 1851, concerning the motion of the asteroid Hygeia (the tenth known asteroid, discovered in 1849). Hind happened to be an expert on the part of the sky in which Ferguson had observed Hygeia, and he noticed that the reference star designated as *k* in Ferguson's charts was nowhere to be found in his own observations of the sky. Hind brought this to the attention of the director of the Naval Observatory, Matthew Fontaine Maury, who confirmed Hind's work. Both Hind and Maury concluded that Ferguson's star *k* was a previously unknown planet. Hind published his own letter in the *Astronomical Journal* in which he estimated that this newly discovered planet was in a 1,600-year orbit at distance of 137 AU from the Sun.

By now, Hind's and Ferguson's planet had disappeared into the rich star fields of Sagittarius, and there was no hope than anyone would be able to find it again, perhaps for decades. Hind's and Ferguson's planet disappeared into the dustbin of history in 1879 when the director of the Hamilton College Observatory in New Jersey, Christian Peters, revisited the problem. Peters recognized that star *k* never existed. Instead, Ferguson had mistakenly noted that he had measured the position of Hygeia relative to wire No. 2 on his micrometer (which had three fixed wires), but the actual measurement was relative to wire No. 1. By inference, there would have been a positional reference star at the position of wire No. 2 when in fact the reference star was at the position of wire No. 1.[1] Since Ferguson's mistake was uncovered only after a quarter century, from 1851 until 1879, the solar system had, for the third time, a ninth planet, the first ninth planet having been one of Jupiter's moons and the second having been the recently demoted asteroid Pallas.

Le Verrier himself believed that a planet existed beyond Neptune; however, he saw no point in trying to find such a planet until several decades of observations of Neptune were in hand. Instead, Le Verrier turned his attention to Mercury and reached the conclusion that "Mercury is without a doubt perturbed in its path by some planet or by a group of asteroids as yet unknown."[2]

The problem with the orbit of Mercury was well established by the mid-nineteenth century. The end points of the major axis of an elliptical orbit connect the perihelion (closest to the Sun) and aphelion (farthest from the Sun) positions. The line running through these two points is known as the *line of apsides*. For Mercury, the line of apsides moved in a regular fashion, precessing—that is, turning in one direction—at a pace of 1.38 seconds of arc per 87.97-day orbit. At this cumulative rate of 574 seconds of arc per century, Mercury's orbit will swing around completely in 226,000 years. In a simple Newtonian two-body calculation involving just the Sun and Mercury, this precession could not occur; however, the combined gravitational influences of all the known planets, including the newly discovered Neptune, but most importantly massive Jupiter, produced a precession of the perihelion (as it came to be called) of 1.28 seconds of arc per orbit (532 seconds of arc per century). The remaining, unexplained one-tenth of a second of arc per orbit is astoundingly small, far too small in fact for nineteenth-century astronomers to have measured in a single orbit. The cumulative effect of one-tenth of a second of arc per orbit is a very difficult to measure 0.43 seconds of arc per year; however, this unexplained precession accumulates to an easily measurable 4.3 seconds of arc in only a decade and 43 seconds of arc per century. Le Verrier's assertion that the orbit of Mercury "without a doubt" has a problem in need of an explanation was incontrovertible. One explanation was that Newton's law of gravity is not quite right; a very slight change in the mathematical expression of Newton's formulation of the law of gravity, astronomers of the mid-nineteenth century realized, might solve the Mercury problem. However, Newton's theory had just experienced its greatest triumph through the discovery of Neptune. To Urbain Le Verrier, the greatest planetary theorist of his time, made famous by his use of Newton's law to predict the existence of Neptune, another explanation was necessary.

Le Verrier concluded, in a letter published in September 1859, that a gravitational influence inside of Mercury's orbit must exist to account for the otherwise unaccounted for amplitude of the precession of the perihelion of Mercury. Perhaps inspired by the newly discovered asteroid belt, he seemed to prefer a swarm of asteroids at 0.3 AU (just inside of the orbit of Mercury, at 0.4 AU) to a single planet as the source of the gravitational perturbations on Mercury's orbit. The mass of the intra-Mercurial planet or the total mass of the swarm of asteroids had to be small enough, and this object or objects had to be close enough to the Sun, so as to not affect the orbits of Venus or any other planets. Yet, the net gravitational influence of the hypothesized intra-Mercurial planet or asteroidal swarm had to be big enough to affect Mercury. Given that the discovery of Neptune resulted from similar calculations by the great Le Verrier, these new calculations carried considerable weight and influence. And since no other explanation seemed reasonable, most astronomers accepted without question, well into the first decade of the twentieth century, that "it is not improbable that there is a considerable quantity of matter circulating around the Sun inside the orbit of Mercury."[3]

Amazingly, the first reported discovery of an inner planet, named Vulcan, actually predated by six months the public announcement by Le Verrier of his prediction of an intra-Mercurial planet. A French doctor and amateur astronomer, Edmond Lescarbault, informed Le Verrier, in a letter written shortly before Christmas 1859, that he had recorded an observation of an intra-Mercurial planet as it passed in front of the Sun. Lescarbault's observations had been made on March 26 of that year, when he claimed to have observed a small, dark spot pass in front of, or transit, the Sun for a total of one hour and seventeen minutes. From the information recorded by Lescarbault, and after having personally interviewed Lescarbault to assess for himself the dependability of the doctor's reported discovery of Vulcan, Le Verrier calculated that Vulcan had an orbital semi-major axis 0.147 AU, an orbital period just over nineteen days seventeen hours, and a mass of one-seventeenth that of Mercury.

Assuming a similar density as the Moon, such an object would still be fairly large, with a diameter just over 60 percent that of the Moon, or 2,100 km, much bigger than the presumed sizes of the asteroids discovered early in the century. However, being at 0.147 AU rather than at 0.3 AU, the mass and location of Vulcan was insufficient, by itself, to account for the anomalous precession of the perihelion of Mercury. Fully twenty planets like Vulcan were needed to solve the Mercury problem; but this swarm of planets (predicted by Le Verrier) and subsequent discovery of Vulcan were just the things, Le Verrier was convinced, to confirm his theory. The third tenth planet (one of the Galilean moons was the first tenth planet; Juno was the second), albeit one with only 26 percent the mass of Earth's moon, allegedly had been found.

Although no one else ever saw Lescarbault's planet, including a professional French astronomer, Emmanuel Liais, who had been studying the Sun from the coast of Brazil on the same day and time as Lescarbault, the influence of Le Verrier, who had become the director of the Observatory of Paris, was profound. For many astronomers, Vulcan and Neptune both existed because Le Verrier's calculations demanded that they exist. Just because Vulcan was hard to spot because of its proximity to the Sun did not change the fact that it must be there. In fact, in 1876 Le Verrier received the Gold Medal from the Royal Astronomical Society for his work in solving the Mercury problem.

As happened immediately after the discoveries of Uranus and Neptune, astronomers looked into old observing records in an attempt to discover historical observations of the same object. Many such historical discoveries were reported, but none of them matched the orbit predicted by Le Verrier for Vulcan. Others, looking toward the future rather than the past, predicted when Vulcan would be observable transiting the face of the Sun or in the dark skies surrounding the Sun during solar eclipses. Unfortunately, Vulcan was hard to spot. As early as March of 1860 and continuing several times every year for two decades, and less often for another three decades, astronomers planned and carried out expeditions to remote locations, trying to relocate the lost planet, but to no avail.

All other observations of Vulcan, dating back to 1762, were dismissed as sunspots. Sunspots are dark spots, or small groups of dark spots, that appear with regularity on the surface of the Sun. They are the manifestation of the tangling of magnetic field lines near the surface of the Sun, which leads to enhanced cooling of certain areas, ultimately making them look darker than the rest of the solar surface.

By the late 1870s, both the innermost (Vulcan) and outermost (Hind's and Ferguson's planet) planets in the solar system seemed on their last legs. Hind's and Ferguson's planet disappeared in 1879, having never been more than a mistaken positional measurement by Ferguson, and then there were but nine planets again. For Vulcan, 1878 "was the watershed year" when an overwhelming amount of evidence would wash away most of the support for the existence of this planet.[4]

The best conditions during which to search for an intra-Mercurial planet occur during total eclipses of the Sun. During those few minutes when the Sun is totally eclipsed, astronomers might hope to detect objects located near the Sun that are much fainter than Mercury. Total solar eclipses are rare, however, and even in the twenty-first century, astronomers have to mount special expeditions to the small, often remote, areas of the Earth's surface affected by the narrow and short ground path of a solar eclipse if they wish to study the near-Sun environment under such conditions.

During a solar eclipse that crossed much of the western United States on July 29, 1878, an array of renowned astronomers and scientists converged near the now–ghost town of First Separation, Wyoming. Some of those involved were determined to prove the existence of Vulcan; others were as equally determined to demonstrate Vulcan was imaginary. James Watson, a believer, an extraordinarily successful asteroid hunter, and the director of the University of Michigan Observatory, reported that during the eclipse he had seen two previously unseen planets. The amateur astronomer Lewis Swift also reported seeing two intra-Mercurial planets while the Sun was in eclipse. Might Le Verrier's intra-Mercurial asteroid swarm exist? On the other hand, as pointed out quite forcefully by Christian Peters, Professor of Astronomy at Hamilton College in New York, the director of the Litchfield Observatory, and a skeptic, the positions of Watson's and Swift's pairs of planets did not match, nor did any of the alleged objects match the positions predicted for the orbit of Vulcan. In addition, the relative positions of the pairs of alleged planets very nearly matched the relative positions of two well-known stars. Perhaps Watson or Swift or both misidentified stars as planets, in part in the haste and excitement of the total solar eclipse and in part due to their enthusiasm for finding such objects. By the end of 1878, few other than Watson and Swift still believed that the First Separation eclipse observations confirmed the existence of the intra-Mercurial swarm of tiny planets.

Le Verrier had died in 1877, and Watson followed him into history in 1880. Although astronomers continued to search for Vulcan during eclipse expeditions to Egypt in 1882 and to Caroline Island in the Pacific Ocean in 1883, Vulcan also was passing into history and by 1890 was less real than suspect. Norman Lockyer, who was a fellow of the Royal Astronomical Society, the discoverer of the element helium, and the founder of the journal *Nature*, reported some skepticism about Vulcan in the American edition of his textbook *Elements of Astronomy*, published in 1881 and revised in 1890. Therein he reports that a ninth planet

—quite small—is suspected to exist, between Mercury and the Sun, only thirteen million miles from the latter, and performing its revolution in about 19¾ days, in an orbit inclined to the ecliptic at an angle of 12°. A French physician, named Lescarbault, claimed to have discovered it crossing the Sun's disk in 1859. The name Vulcan was assigned to it.

Other observers have, at different times, seen spots of a planetary character rapidly cross the disk of the Sun, which may turn out to have been transits of Vulcan; but up to the present time we can only say that the existence of such a planet is suspected—it is not proved.[5]

A decade later, Charles A. Young, Professor of Astronomy at Princeton University, wrote the following in his 1901 revision of his own highly regarded textbook, also titled *Elements of Astronomy*:

It is not improbable that there is a considerable quantity of matter circulating around the Sun inside the orbit of Mercury. This has been believed to be indicated by an otherwise unexplained advance of the perihelion of its orbit. It has been somewhat persistently supposed that this intra-Mercurial matter is concentrated into one, or possibly two, planets of considerable size, and such a planet has several times been reported as discovered, and has even been named "Vulcan." We can only say here that the supposed discoveries have never been confirmed, and the careful observations of total solar eclipses during the past ten years make it practically certain that there is no "Vulcan."[6]

By the mid-1890s, even the theoretical work of Le Verrier that had been the original impetus for the search for Vulcan had been called into question. Simon Newcomb, the most celebrated celestial mechanician of the late nineteenth century, had recalculated all the orbits of the four inner planets and, in doing so, corrected some fairly significant errors and inconsistencies in the work of Le Verrier. He concluded that any intra-Mercurial swarm that might exist also would affect the orbit of Venus. Consequently, since Mercury's orbit definitely was affected by something while Venus's orbit was unaffected, Newcomb and others began to tinker with possible corrections to Newton's law of gravity.

Still, without some explanation based on newly understood laws of physics, others continued to try to solve the Mercury problem within the realm of the laws of physics as then understood. Simply put, the search for Vulcan continued though with less enthusiasm. One eclipse expedition took astronomers to the island of Sumatra in 1901, but no evidence of an intra-Mercurial planet was found. Perhaps Vulcan, by bad luck for the observers, was hidden behind the Sun during the 6½ minute duration of the eclipse. Even as late as 1908, special expeditions were mounted to search for Vulcan or any other intra-Mercurial planet, but Vulcan and its companions were gone, never to be seen again.

Vulcan, of course, did not and does not exist. The precession of the perihelion of Mercury is now explained as a consequence of Albert Einstein's general theory of relativity, which can be understood as a small—though revolutionary—correction to Newton's law of gravity. Near a small, fairly low-mass object like the Earth, the relativistic corrections to Newton's law of gravity are extremely small; however, in the near environment of a massive object like the Sun, Newton's law of gravity fails more dramatically. Thus, Einstein would conclude in 1916, the precession of Mercury's perihelion is a consequence of the large curvature of space-time near the Sun.

The solar system, which for most of the second half of the nineteenth century was believed by many astronomers to have had nine or even ten planets, once again had only eight.

Easy come, easy go.

10

Pluto, the Fourth Ninth Planet

O

By 1877, when the American astronomer David Peck Todd began the first serious observational search for a planet more distant than Neptune, the solar system was once again down to nine planets (Hind's planet was now known to be an error, but Vulcan retained many supporters until well past 1900). Todd used the possible unexplained differences, called residuals, in the observed versus predicted positions of Uranus (not Neptune!) to posit a trans-Neptunian planet 52 AU from the Sun. These residuals would be the differences not explained by the presence of Neptune. He even calculated the longitude of the hypothesized tenth planet, but despite six months of searching, he came up empty.[1]

A year later, the Frenchman Camille Flammarion used a new idea to predict the existence of two trans-Neptunian planets, one 100 AU from the Sun and the other an incredible 300 AU from the Sun. Flammarion noted that several comets seemed to come from the same distant place in the outer solar system; he suggested that this clustering of the aphelia of comets was due to the existence of a distant planet at that position. Flammarion's planets were never found.[2]

In 1900, the Danish astronomer Hans-Emil Lau used the residuals in the orbit of Uranus, now covering two hundred years from 1690 to 1895, and determined, like Flammarion, that two trans-Neptunian planets existed. Lau's planets were closer, though, at distances of only 46.6 and 70.7 AU. Lau also assigned masses to his planets, 9 and 47.2 times the mass of the Earth, respectively. Given that Lau's solution was that of an inverse problem, he could determine two heliocentric longitudes, rather than only one, for each of his planets, with the two solutions for the longitude being on opposite sides of the Sun. Lau's planets were never found on either side of the Sun.[3]

By now, the playing field had become crowded. Between 1901 and 1903, the Europeans Gabriel Dallet (47 AU), Theodore Grigull (50.6 AU) and Vicomte du Ligondes (about 50 AU) all made calculations in rough agreement, not just for the size of the missing planet's orbit but also for the longitude. In 1902, a Russian, Alexander Garnowsky, predicted the existence of four trans-Neptunian planets. In 1904, the American Thomas Jefferson Jackson See determined that three trans-Neptunian planets were to be found at 42.25, 56, and 72 AU. None of the planets predicted by Dallet, Grigull, Ligondes, Garnowsky, and See were ever found.[4]

This game had come to resemble the story of the boy who cried wolf. The predictions regarding trans-Neptunian planets were treated skeptically by most of the astronomy community, and this skepticism was reinforced by the disagreements in predictions between the advocates and the failures of any planet hunters to find any of these planets. Then, in one of the more prescient and under-appreciated aspects of this enterprise, the Frenchman Jean Baptiste Aimable Gaillot reviewed the orbits of Uranus and Neptune calculated new orbits and position tables for these two planets, and concluded that no evidence of a trans-Neptunian planet existed in the extant body of data for Uranus and Neptune. In other words, there were no residuals to be explained. The orbit of Uranus was well explained by the presence of Neptune and the other known planets, while Neptune's orbit was completely understood by the presence of Uranus and the other known planets.[5]

PICKERING AND LOWELL

The work of Gaillot put the issue of the existence of a trans-Neptunian planet to rest for most of the solar system astronomy community, but not for all: two American astronomers refused to let the idea of Planet X die. William H. Pickering, of Harvard, the ambitious brother of Edward Charles Pickering, director of the Harvard College Observatory, spent much of the next two decades making graphical predictions for trans-Neptunian planets, in particular his Planet O, based on his analysis of residuals in the orbit of Uranus. And Percival Lowell, a rich gentleman from Boston, trained in mathematics and a self-taught astronomer who built himself an observatory in Flagstaff, Arizona, dedicated to observing Mars and proving that Martians existed, dedicated much of the last eight years of his life to a fruitless search for Planet X.

Pickering first announced the alleged existence of Planet O in 1908. Planet O, he claimed, was twice as massive as the Earth and could be found at a distance of 51.9 AU from the Sun. In 1911, Pickering announced the probable existence of three more planets beyond Neptune: Planet P at 123 AU, Planet Q at 875 AU (and a mass twenty thousand times that of the Earth, or sixty-four times the mass of Jupiter), and Planet R at 6,250 AU (and a mass ten thousand times that of the Earth, or thirty-two times the mass of Jupiter). Over the next two decades, Pickering also postulated the existence of three other trans-Jovian planets, called Planets S, T, and U. In 1931, Planet S emerged from Pickering's calculations as a 5-Earth-mass object at 48.3 AU, and Pickering calculated that Planet T would appear at 32.8 AU. A year later, Planet U showed up in Pickering's theory as a small, 0.045-Earth-mass object just beyond Jupiter, at 5.79 AU. The orbit of Planet U permitted it to cross back and forth across that of Jupiter.

Ultimately, Pickering would so completely revise the parameters for Planets O and P that author William Graves Hoyt, in his book *Planets X and Pluto*, considers them altogether different planets than the original Planets O and P, giving Pickering a grand total of nine planets predicted to exist in the outer solar system.

In 1919, Pickering published a revision of his orbital parameters for Planet O. Planet O, in the 1919 incarnation, would be found at 55.1 AU, still with a mass of two Earth masses. In 1928, he reduced the size of the orbit of Planet P from 123 AU to 67.7 AU and assigned it a mass of 20 Earth masses. In 1931, he would move Planet

P back out again, to 75.5 AU, and increase its mass to 50 Earth masses. As for Planet O, in 1928 Pickering reduced its mass to 0.75 Earth masses. Pickering's 1928 orbit also predicted an eccentricity for Planet O such that Planet O crossed the orbit of Neptune. Of most interest for consideration later is the 1930 epoch heliocentric longitude of 102.6 degrees predicted by Pickering for Planet O, in 1919. Over two decades, from 1908 through the 1920s, Pickering searched for and encouraged others to search for his Planet O, but Planet O was rarely searched for and never found, nor were planets $P, Q, R, S, T,$ or U.

Percival Lowell, born into New England affluence and whose brother was for twenty-four years the president of Harvard University, distinguished himself in studying mathematical astronomy as an undergraduate at Harvard. Thereafter, he embarked on a career in business and as a diplomat until, in 1893 at the age of thirty-eight, he rediscovered astronomy. Specifically, he (mis)learned enough about late nineteenth century studies of Mars to convince himself that life existed on that planet. He financed the construction of his own observatory in Flagstaff and hired a staff to help him in his research. Only in 1905 did he begin his quest to find the yet unfound trans-Neptunian planet, which he would call the search for Planet X. He would continue his search—using both calculations and telescopic surveys—until his death in 1916.

Lowell himself recognized the difficulty of the problem he was trying to solve. In his 1915 paper "Memoir on a Trans-Neptunian Planet," he pointed out, "We cannot use Neptune as a finger-post to a trans-Neptunian as Uranus was used for Neptune because we do not possess observations of Neptune far enough back."[6] (In fact, Galileo had unknowingly observed Neptune, on at least two and possibly as many as four occasions, from late December 1612 into early 1613. These observations might have helped Lowell enormously; however, Galileo's observations of Neptune were not recognized until 1979, in work by Stillman Drake and Charles Kowal.) Lowell further noted that in 1845, when Adams and Le Verrier used the residuals in the orbit of Uranus to predict the existence of Neptune, the difference between the predicted and observed positions

Figure 10.1: Percival Lowell. (Image used by permission of Lowell Observatory)

of Uranus sometimes amounted to 133 seconds of arc. In contrast, after accounting for the gravitational influence of Neptune on Uranus, the remaining residuals in the orbit of Uranus never exceeded 4.5 seconds of arc.

Lowell used the tried-and-true mathematical methods of Adams and Le Verrier—examining the apparent residuals in the orbit of Uranus—and an assumption that the semi-major axis of Planet X would be close to 47.5 AU, to work out his initial prediction for the location of Planet X. He apparently started his calculations with an assumption akin to the Titius-Bode rule: the relative distances of all planets were determined by a not-yet-understood physical principle that forced the ratios of orbital periods of neighboring planets to be close to ratios of small integers. No physics supported such an assumption, though many astrophysicists even today believe that some physical process, not yet identified, controlled the spacing of the planets as they formed.

Lowell asserted that because the ratio of the orbital period of Neptune (164.8 years) to that of Uranus (84.0 years) is very nearly two, the ratio of the orbital period of Planet X to that of Neptune also should be nearly two. According to Kepler's third law, if Planet X were in an orbit with a semi-major axis of 47.5 AU, it would have an orbital period of 328 years, which is almost twice the orbital period of Neptune.

One can quickly evaluate the utility (or lack thereof) of this supposed physical principle. For example, the ratio of the orbital periods of Neptune and Uranus, 1.96, is only 2 percent away from the ratio of small integers 2:1; the ratio of the orbital period of Uranus to that of Saturn (29.46 years) is 2.85, which is off only by 5 percent from the ratio of the small integers 3:1; the ratio of the orbital period of Saturn to that of Jupiter (11.86 years) is 2.484, which is fairly close to the ratio of small integers 5:2; the ratio of the orbital period of Mars (686.98 days) to that of the Earth (365.25 days) is 1.881, which is in error by 6 percent from the ratio of small integers 2:1; the ratio of the orbital period of Earth to that of Venus (224.70 days) is 1.626, which is off by 8 percent from the ratio of the small integers 3:2; and the ratio of the orbital period of Venus to that of Mercury (87.97 days) is 2.554, which is off by 5 percent from the ratio of the small integers 5:2. Lowell's rule would seem to be comparable in accuracy to, though if anything a bit worse than, the failed Titius-Bode rule. But to Lowell, this rule was an important criterion for guiding his initial calculations and predictions. In 1913, Lowell predicted that Planet X should have a mass of 6.6 Earth masses and a semi-major axis of 47.5 AU. Apparently, he soon discarded his own rule, as his 1914 calculations reduced the semi-major axis to either 43.0 or 44.7 AU.

From 1906 through 1916, Lowell directed a photographic search of the sky for Planet X. Amazingly, two of the photographic plates obtained at Lowell Observatory in 1915 contain images of Pluto, but because Lowell was looking for a much brighter object, his sought-after Planet X slipped through his fingers.

CLYDE TOMBAUGH DISCOVERS PLUTO

When Lowell died in 1916, his hypothesized Planet X remained hidden in the darkness of the outer solar system. It would take the better part of a decade after his death for the finances of Lowell Observatory to get sorted out, but by 1927 the observatory, newly under the leadership of Percival Lowell's nephew Roger Lowell Putnam, decided to launch a new search for Lowell's planet and to hire an amateur astronomer to assist in the program. Fortuitously, the young amateur astronomer Clyde William Tombaugh had just initiated a correspondence with V. M. Slipher, director of Lowell Observatory, concerning Tombaugh's drawings of Mars and Jupiter. Within months, Tombaugh was on a train to Flagstaff.

Tombaugh was a farm boy from Kansas, self-educated in astronomy, who arrived in Flagstaff in February 1929 at age twenty-three, having no idea what his role would be at the Flagstaff observatory. On April 6, 1929, Tombaugh began his search for Planet X.

His start was inauspicious. Visiting astronomers advised him that he was wasting his time and that the search techniques devised by his professional colleagues would not work. Then, in October, the stock market crashed, putting a huge crimp in the cash flow of the observatory. Tombaugh's work was exceedingly hard and tedious and required extraordinary dedication and patience. He spent his nights in a cold telescope dome, carefully acquiring long exposures on 14 inch by 17 inch photographic plates of selected regions of the sky. Each region of the sky had to be observed multiple times. Then, carefully comparing the positions of the more than three hundred

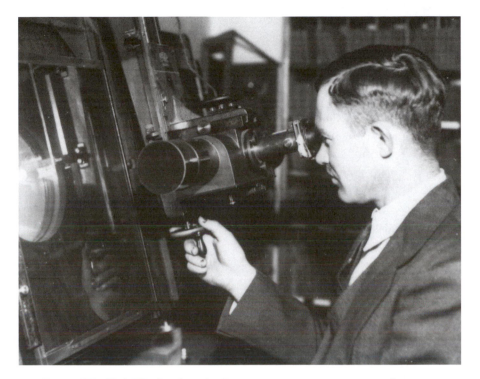

Figure 10.2: Clyde Tombaugh at the Zeiss comparator, comparing two photographs for moving objects. (Image used by permission of Lowell Observatory)

thousand stars on each plate with their positions on the same image of the sky obtained a few days later, Tombaugh looked for evidence of a slowly moving object. His examination of only a few square inches of the photographic plates was a full day's work.

Nevertheless, on February 18, 1930, Clyde Tombaugh found Pluto while carefully examining images obtained a month earlier, on January 23 and 29. Having previously counted in and then out both Jupiter's moon Europa and the asteroid Pallas as the ninth discovered planet, and having lost intra-Mercurial Vulcan as the ninth planet about the same time that Percival Lowell became interested in finding Planet X, the solar system had a ninth planet for the fourth time.

Lowell Observatory publicly announced Tombaugh's discovery to the world on March 13, a date that was carefully selected because it happened to be the 75th anniversary of Percival Lowell's birth and the 149th anniversary of the discovery of Uranus. In the Observation Circular, entitled "The Discovery of a Solar System Body Apparently Trans-Neptunian" because the name Pluto had not been selected yet, Lowell Observatory director V. M. Slipher noted, "In its apparent path and in its rate of motion it conforms closely to the expected behavior of a Trans-Neptunian body, at about Lowell's predicted distance . . . its position and distance appear to fit only those of an object beyond Neptune, and one apparently fulfilling Lowell's theoretical findings."[7]

Nevertheless, almost from the date on which the discovery of Pluto was announced, Pluto generated controversy: it was almost ten times fainter than Lowell had predicted for Planet X, and this implied that it was significantly smaller and less massive than predicted by Lowell for Planet X. Quietly, E. C. Slipher of the Lowell Observatory staff began making measurements to determine the size and thereby to infer the mass of Pluto. He concluded that although a disk as small as 0.25 seconds of arc should be detectable with the Lowell Observatory 61 cm telescope used by Tombaugh, an object twice that size might look pointlike or starlike. By inference, Pluto could be bigger than it looks.

Within months, debates over the nature of Pluto began. Was it Lowell's Planet X or Pickering's Planet O? Was it neither? Could it be a trans-Neptunian comet? Tombaugh himself noted, in his reminiscences fifty years later, that "in the course of several weeks, the feeling grew that the real Planet X was not yet found."[8] Pluto's orbit was difficult to calculate, but those doing the work rapidly converged on a solution with a semi-major axis close to 40 AU and an orbit more highly inclined to the ecliptic plane than the orbit of any other planet. By the middle of 1930, Pluto's orbit had been refined to a semi-major axis of 39.5 AU, an eccentricity of 0.248, and an inclination to the ecliptic plane of 17.1 degrees. For comparison, the next most inclined orbit is that of Mercury, at 7.0 degrees; Mercury also has the next most eccentric orbit, with a value of 0.206. Clearly, Pluto was at the extreme edge, literally and in the shape and tilt of its orbit, of the known planetary system.

The mass of Pluto was smaller than expected. Lowell had predicted 6.6 Earth masses; Pickering had predicted 2.0 and then 0.75 Earth masses. Given the angular size and distance of an object, the mass of that object can be estimated fairly well from simple geometric principles. Pluto, however, in contrast to all the other planets, was too small for even the biggest telescopes to resolve. That is, it looked like a point of light, just like a star, showing no disk. Obviously, it was small and geometric smallness implied physical smallness and a mass well below Lowell's predictions. How far below the predicted mass was now the tantalizing question.

From the brightness of Pluto, combined with an assumption for Pluto's density (i.e., its composition) and an estimate of how effectively the surface of Pluto reflects sunlight, one can estimate the size and then the mass of Pluto, even without directly measuring the size. A big, highly reflective object (e.g., aluminum) will reflect more light than a small object made of the same material; however, a large object made of poorly reflecting material (e.g., charcoal) may reflect the same total amount of light as a small, highly reflecting object. An astronomer can directly measure the amount of light received from an object; however, if the object is so small that its size cannot be measured directly, the astronomer has to guess how reflective the object is. Such guesses are usually fairly good, and at worst will be off by a factor of a few. Such was the case for Pluto. Using the measured brightness of Pluto and a value estimated for Pluto's reflectivity, astronomers were able to estimate the physical size of Pluto. Then, by making assumptions about the likely density of Pluto (is it made of rock? is it half rock and half water ice?), astronomers could calculate Pluto's mass. Since the density estimate could not possibly be wrong by more than a factor of three (the density of water ice and rock are about 1 and 3 gm per cubic centimeter, respectively), a final estimate for the mass of Pluto should be at worst ten or twenty times too small or too big. Given the need to make two educated guesses (the reflectivity of Pluto's surface, the internal composition) in order to calculate the mass of Pluto, we should expect different astronomers to get slightly different answers—and they did.

In 1930, Ernest William Brown, of Yale University, estimated a mass of 1 Earth mass for Pluto based on the imperceptibility of Pluto's gravitational perturbations on the outer planets. Also in 1930 and using the same method as Brown, the English astronomer A.C.D. Crommelin suggested a mass of only 0.11 Earth masses, while in 1931, Ernest Clare Bower, of Lick Observatory, conservatively estimated Pluto's mass to be less than 0.7 Earth masses, but suggested a most probable mass at 0.1 Earth masses. Not surprisingly, these three astronomers obtained different answers; however, their answers were also impressively consistent, putting Pluto's mass in the range of 0.1 to 1 Earth mass.

Immediately, these calculations raised a red flag. The reason that Lowell and Pickering predicted the existence of Pluto was the set of unexplained residuals they claimed existed in the orbits of Uranus and Neptune. If Pluto is less massive than predicted, then the gravitational effect of Pluto on Uranus and on Neptune also must be less than that predicted. Since Pluto was more than nine times, and most probably sixty times, less massive than estimated by Lowell, the gravitational influence of Pluto on anything would be more than nine times less, and probably more than sixty times less, than originally anticipated. Therefore, Pluto was unlikely to have produced the effects on the orbit of Uranus that led Lowell to predict its existence in the first place.

Yet, Pluto was found.

Furthermore, Pluto was found fairly close to the position predicted for it by Lowell. Actually, one should say that Pluto was found fairly close to one of the two possible positions predicted for it by Lowell. Specifically, when discovered, Pluto was 5.8 degrees away from one of two heliocentric longitudes predicted for it by Lowell for that date (remember that the solution to an inverse problem typically has more than one reasonable answer). Amazingly, Pluto was only 5.9 degrees from one of the many positions predicted for Planet O by Pickering, with Pickering's and Lowell's predictions differing by only 0.1 degrees.

"Fairly close," like beauty, is in the eye of the beholder. Johann Galle located Neptune fifty-five minutes of arc, or just less than one degree, from the position predicted for it by Urbain Le Verrier. In other words, Le Verrier's prediction for the position of Neptune was six times better than the positions predicted for Pluto by Lowell and Pickering. Given that Lowell predicted two positions for Planet X, on opposite ends of the sky, the range on the ecliptic within six degrees east or west of both of those positions encompasses twenty-four degrees, or 7 percent, of the ecliptic. In fact, Tombaugh had already searched star fields in Aquarius, Pisces, Aries, and Taurus before he found Pluto in Gemini. He had searched more than one-third of the zodiac, the set of constellations that enclose the ecliptic, prior to stumbling upon Pluto. Was the discovery of Pluto due to Lowell's good prediction, dumb luck, or the diligence and perseverance of Tombaugh?

SHRINKING PLUTO

Pluto's mass soon began to shrink further and then grew again. In 1931, Seth B. Nicholson, of Mount Wilson Observatory, assigned a mass to Pluto in the range from 0.69 to 1.19 Earth masses, based on the absence of any apparent gravitational influence of Pluto on the other planets, and Pickering made his own graphical guesstimate of 0.71 Earth masses, in marvelous agreement with his prediction for Planet O. Ernest Bower made new estimates of 0.1 to 0.3 Earth masses (1933) and 0.06 to 0.09 Earth masses (1934). A decade would pass before the Pluto mass game began again in earnest. In 1942, Lloyd R. Wylie, of the U.S. Naval Observatory, calculated a mass of 0.91 Earth masses based on the strength of the gravitational tugs he assumed Pluto must exert on Uranus and Neptune to explain their discordant positions. Then in 1949, Gerald R. Kuiper, working at the McDonald Observatory in west Texas, measured the angular size of Pluto to be no more than four-tenths of a second of arc, giving Pluto a maximum diameter of 10,300 km. Assuming a reasonable density for Pluto, Kuiper used his size measurement to estimate a maximum mass for Pluto of 0.8 Earth masses. In March, 1950, Kuiper used the Palomar Observatory, in Southern California, and determined that the angular size of Pluto was only 0.23 seconds of arc. He further concluded that this measurement was "comparatively simple" and that he had measured, without a doubt, the real size of Pluto, not an upper limit. With this measurement in hand, and again assuming a density for Pluto, he concluded that the diameter of Pluto was 0.46 earth diameters, or 5,800 km, and the mass 0.1 Earth masses. The last calculation that yielded a large mass for Pluto was made in 1950, when Dirk Brouwer, of Yale University, gave an estimate of 0.91 Earth masses, using the same general method as Wylie. Thereafter, Pluto stayed small.

In 1965, a team at the U.S. Naval Observatory in Flagstaff, a neighbor of Lowell Observatory, determined that Pluto's angular size was less than 0.25 seconds of arc and its mass less than 0.14 Earth masses. In 1968, using theoretical arguments having to do again with residuals in the motion of Neptune, R. L. Duncombe, W. J. Klepczinski, and P. K. Seidelmann of the U.S. Naval Observatory reduced the mass of Pluto to 0.18 Earth masses. In 1970, Dennis Rawlins produced his own estimate, in agreement with the 0.18 Earth mass estimate of Duncombe, Klepczinski, and Seidelmann. In 1970, Duncombe and his team reduced the mass of Pluto to 0.17 Earth masses, again based on a theoretical reanalysis of the residuals in Neptune's orbit. In 1971, J.G. Williams of Caltech and G. S. Benson of the University of California–Los Angeles also argued for a mass of 0.17 Earth masses, while Seidelmann and his U.S. Naval Observatory colleagues further reduced Pluto's mass to 0.11 Earth masses. After forty years, astronomers had returned Pluto's mass to its 1931 starting point.

Further shrinkage would come. P. Baldi and M. Caputo, in 1974, shrunk Pluto into the range 0.004 to 0.014 Earth masses, making it comparable to or smaller than the Moon. Then in 1976, Dale Cruikshank, Carl Pilcher, and David Morrison of the University of Hawaii were able to determine that Pluto was better at reflecting light than anyone previously had thought; rather than being dark, like coal, Pluto is bright, like freshly made ice. With a surface that more efficiently reflects light back to Earth, Pluto could be smaller

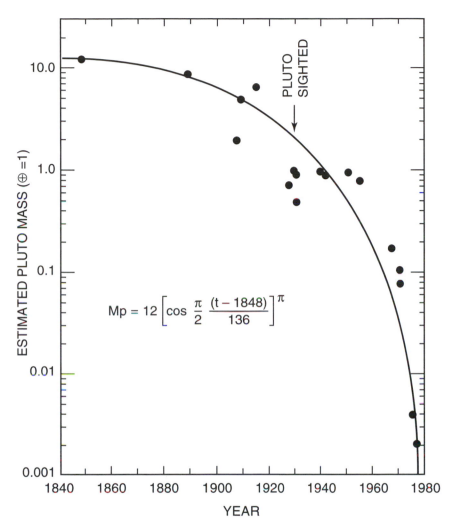

Figure 10.3: Tongue-in-cheek graph depicting the change in the measured mass of Pluto from 1848 through 1978 (Eos 61 [44] [October 1980]: 690). A. J. Dessler and C. T. Russell teased that if we took these data seriously, i.e., if Pluto's mass were really shrinking, then Pluto's mass would become negative in 1984. In this graph, Pluto's mass is measured in units of Earth masses. (Graph courtesy of Dessler and Russell)

and reflect the same total amount of light. When they estimated Pluto's mass, assuming as had Kuiper a reasonable density for Pluto, Cruikshank and his colleagues had shrunk the mass of Pluto to only 0.003 Earth masses, less than one three-hundredths of the mass of the Earth and barely one-quarter the mass of the Moon.

The endgame was about to begin.

On June 22, 1978, James Christy of the U.S. Naval Observatory, in analyzing photographic plates obtained for him by the U.S. Naval Observatory staff astronomer Anthony Hewitt in April and May of that year, discovered that Pluto had a moon, which he named Charon. From Kepler's laws, the orbital period of a moon combined with the size of the orbit immediately gives the total mass of the planet plus the mass of the moon. Christy and his colleague Robert Harrington used this method to calculate the total combined system mass; they obtained an answer of 0.0017 Earth masses, which in 1980 they revised to 0.0023 Earth masses (with an error of about 15 percent). Since Charon was much fainter than Pluto, it almost certainly was much smaller and less massive than Pluto. Christy and Harrington guessed that 90 to 95 percent of the total mass of the Pluto-plus-Charon system would be the mass of Pluto. With Charon discovered and its orbit known, Pluto's mass had shrunk to very nearly 0.002 Earth masses.

In 2003, C. B. Olkin, L. H. Wasserman, and O. G. Franz, from Lowell Observatory, used the Hubble Space Telescope to refine the mass ratio between Pluto and Charon. During the 1990s, this ratio had variously been determined to be between about 6 and 12. The result derived from the Hubble observations was 8.2, with an accuracy of 6 percent. Using this mass ratio, Olkin, Wasserman, and Franz found that the mass for Pluto is 0.0022 Earth masses, to an accuracy of 5 percent, making Pluto about 450 times less massive than the Earth.

Pluto never quite disappeared, but it came close. More important, by 1978 it was abundantly clear that Pluto was neither Planet X nor Planet O; Pluto was not capable of exerting any significant gravitational influences on the orbits of either Uranus or Neptune. In fact, despite the great distances between the planets,

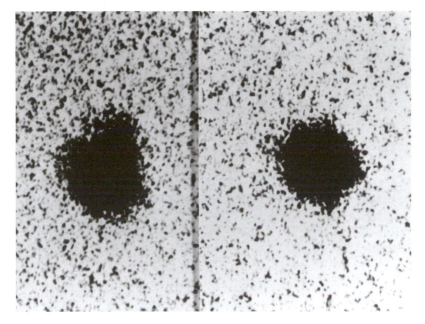

Figure 10.4: Discovery images of Charon, taken with the 1.55-meter (61-inch) telescope at the U.S. Naval Observatory Flagstaff Station. *Left,* elongated image of Pluto. The elongation is due to the appearance of Charon to the upper right. *Right,* Pluto undistorted by Charon because the moon is either in front of or behind Pluto. (Image used by permission of U.S. Naval Observatory)

the Earth has greater perturbative effects on the orbits of both Uranus and Neptune than does Pluto. The existence of Pluto could in no way have ever been predicted by Lowell or Pickering or anyone else, based on an examination of the residuals in the orbits of Uranus or Neptune or based on the apparent clustering of the aphelia of comets. *The existence of Pluto, and certainly the actual location of Pluto, simply could not have been predicted.*

Yet, Pluto was discovered.

Nearly a century after Jean Gaillot had first dismissed any and all claims for the existence of a trans-Neptunian planet that were made on the basis of supposed residuals in the orbits of either Uranus or Neptune, E. Myles Standish Jr., working at the Jet Propulsion Laboratory in California, definitively put to rest all claims that unexplained residuals existed. Standish, noting that even some modern authors have continued to make claims for a *trans-Pluto* planet based on such residuals, demonstrated that no such residuals exist.[9] Standish took advantage of the fact that by the early 1990s, NASA spacecraft had visited every planet in the solar system except Pluto and, consequently, the exact positions, orbits, and masses of the planets and all of their large moons were finally known to great precision. The corrections to the masses of the Jovian planets, based on *Voyager* spacecraft encounter data, are especially important. For Neptune, the mass correction was nearly 0.5 percent of the previously accepted value.

Standish was able to show that all of the extant observations of Uranus and Neptune can be explained without any residuals above the observational errors associated with each set of observations. He acknowledged that systematic errors remain in the residuals for all the planets, but he showed that all of these are well explained by well-known problems with the optical data. As Standish explained, "[W]ith all measurements, there is a limit to the accuracy beyond which one cannot expect to extract valid information. There are many cases where that limit has been exceeded; Planet X has surely been such a case."[10]

In the end, we are forced to conclude that no significant residuals ever existed in the Uranian data after the discovery of Neptune, and no meaningful residuals ever existed in the Neptunian data. Whether the object was called Planet O, Planet X, or Pluto, *no mathematical techniques could ever have been applied to our knowledge of the orbits of the outer planets that would have predicted the existence of a trans-Neptunian planet.*

Yet Pluto was discovered.

Pickering and Lowell were both wrong. Though no one would know the reasons for their flawed predictions for more than half a century after the discovery of Pluto, we now know that their predictions were based on attributing a greater accuracy to the data they had in hand than those data merited and in interpreting those data with a subtle bias that they did not recognize they had. Their calculations and conclusions were overly influenced by their expectations; they were not sufficiently neutral in assessing the quality of their own science.

The bottom line is that Lowell Observatory should never have hired a telescope assistant *whom the director intended to assign the task of searching for Planet* X. Clyde Tombaugh should never have been hired *for the specific project of finding Percival Lowell's planet.* The job of looking for Planet X *never should have been assigned to anyone.* Tombaugh never should have made the photographic observations of Gemini on January 23 and 29, 1930. Yet, despite the incorrect predictions and this long chain of events that should not have occurred, Tombaugh was hired, and within a few months Tombaugh found Pluto, an object with several thousand times less mass than the object Tombaugh thought he was looking for, "only" 5.8 degrees from one of the two spots predicted for Planet X/O by Lowell/Pickering.

Simply put, the object Tombaugh found was not the object he was looking for, and the object he was looking for does not exist. So why did Tombaugh find the object we call Pluto relatively near the location predicted for Planet X? And why, despite another thirteen years of prodigious effort searching for additional trans-Neptunian planets, during which he examined images of ninety million celestial objects, spent seven thousand hours blinking pairs of photographic plates, discovered hundreds of asteroids and variable stars and dozens of new clusters of galaxies and one supercluster of galaxies, did Tombaugh fail to find any other objects in the outer solar system?

11

Hidden Secrets of the Outer Solar System

O

On November 1, 1977, Charles Kowal, working with a pair of images obtained on October 18 and 19 with the 1.22 m telescope on Palomar Mountain in Southern California, discovered a new object in the outer solar system. Using the same technique as that employed in 1930 by Clyde Tombaugh to discover Pluto, Kowal saw that his object moved about three minutes of arc through the background stars over the course of twenty-four hours. An additional observation obtained by Tom Gehrels on the same telescope a week *before* Kowal obtained his data on October 18, and other prediscovery photographs dating back to 1895, permitted Brian Marsden, director of the Minor Planet Center in Cambridge, Massachusetts, to calculate the orbit of this new object. Kowal's object has a semi-major axis of 13.7 AU and an orbital eccentricity of 0.38, making it an odd and extraordinary find. This semi-major axis placed the newly discovered object far beyond the outer edge of the asteroid belt, which is inside of 5 AU but well inside the Oort comet cloud, the presumed reservoir for comets that lies well outside of the orbit of Pluto. This netherland between the Oort Cloud and the asteroid belt was thought to be empty except for the giant planets and Pluto; yet, Kowal had found an object in that vast emptiness.

Was this object a comet that had not found its way into the inner solar system or an asteroid kicked outward from the asteroid belt? Since the object showed no indication of cometlike features, astronomers assumed it was an asteroid. Thus, Kowal's object was assigned a provisional asteroidal designation 1977 UB,[1] then the asteroidal number 2060, and Kowal, given the privilege of naming his object, chose the name Chiron, the wisest, gentlest, and most learned of the half-man-half-horse centaurs of Greek mythology.[2]

In 1980 and 1981, J. Degewij, Dale Cruikshank, R. W. Capps, and William Hartmann made measurements of the average brightness and colors of (2060) Chiron and concluded that water ice was not the dominant surface material. In other words, Chiron did not resemble a comet. Instead, they hypothesized that Chiron had a dark, stony surface, similar to that found on some moons in the outer solar system. Seven years later, D. J. Tholen, Hartmann, and Cruikshank discovered that Chiron was two and a half times brighter than expected; they suggested that Chiron had experienced an outburst similar to a volcanic eruption. Such behavior is expected for comets, but not for stony, asteroid-like bodies. Finally, in 1989 Karen Meech and Michael Belton observed a small coma, like those associated with comets, surrounding Chiron. More than a decade after Kowal's discovery, these observations settled the debate about the nature of Chiron: only an icy, cometlike body that warms up as it approaches the Sun can experience outbursts and develop a coma. Yet Chiron is too big to be a normal comet.

Chiron is an odd object, neither fully cometlike nor asteroidal. Every now and then, Chiron experiences an outburst of gas from its surface, as might a comet. Each outburst produces a coma that can stretch 1 million km (three times the distance from the Earth to the Moon) from Chiron. Such outbursts can be triggered only by the pressurized eruptions of extremely volatile gases such as carbon monoxide and methane. These gases, in turn, are produced by solar heating of Chiron's surface, which in turn heats and vaporizes carbon monoxide and methane ices just below Chiron's surface. When the crust becomes thin and weak enough, the heated gases erupt into space. Thus planetary astronomers concluded, without a doubt, that the composition of Chiron is similar to that of comets, which are large chunks of frozen gas. Big comets, however, are nevertheless small objects, perhaps 25 km in diameter, and consequently are faint objects when seen in reflected sunlight at a distance of 13 AU from the Sun. Brighter than a comet, Chiron also is much bigger than a comet. In fact, Chiron has a diameter of between 148 and 208 km, making it more like an asteroid and comparable in size to some of the larger asteroids. Asteroids, however, are not snowballs made of frozen water and carbon dioxide; they are rocks, made primarily of silicates and iron. Thus we have Chiron, the object made of cometary materials that behaves like a comet but has the size of and orbits like an asteroid. In naming this object after a half-man-half-horse centaur, Kowal had indeed made a very wise and meaningful choice.

Given its high eccentricity, Chiron is on an unstable orbit. At perihelion, Chiron reaches 8.4 AU, while at aphelion, it moves as far from the Sun as 18.8 AU. Such an orbit is extremely unstable against perturbations caused by the gravitational influences of the giant planets: Chiron can come as close as 0.4 AU to Uranus and as close as 3.2 AU to Jupiter, and it *crosses* Saturn's orbit and thus could collide with Saturn or an object in the Saturn moon or ring system. Such near encounters allow the giant planets to tug at Chiron and continually change its orbit; eventually, these small changes will almost certainly cause Chiron to come so close to a giant planet that it will experience one or more dramatic close encounters. Within a few million years, one of these close gravitational encounters with one or more of the giant planets undoubtedly will toss Chiron out of its current orbit. Chiron's new orbit may guide it into the inner solar system, perhaps directly into the Sun or onto a collision course with Earth or Jupiter or Saturn, or the new orbit may cast Chiron right out of the solar system into interstellar space. Chiron's future is nearly a mirror image of its past: it must have experienced a series of close gravitational encounters within the last few million years that tossed it into its current orbit. Clearly, Chiron's lifetime as a Saturn crosser is short relative to the 4.56-billion-year age of the solar system; therefore, Chiron must have come from somewhere else prior to finding itself in its current location in the solar system. But from where could Chiron have come?

The first hint of *from where* had emerged three decades earlier, in the work of the Irishman Kenneth Essex Edgeworth and the

Dutchman Gerald Kuiper. Edgeworth speculated, in the 1940s, that at great distances from the Sun, the processes that would contribute toward making planets would operate much more slowly and likely would fail prior to completion, leaving many small planets rather than one large planet in the trans-Neptunian zone of the solar system. In a 1949 paper, he suggested the possibility that such a reservoir might still exist and might be the nearly stable home of comets, prior to those objects being perturbed into the inner solar system where we see them.

Kuiper, using different arguments, came to a similar conclusion. Kuiper wondered why the surface mass density of the solar system ended so abruptly at Neptune. If one could take the mass of Jupiter and spread it out in a ring around the Sun of some reasonable width (perhaps from 4 to 6 AU), one would obtain an estimate of the *surface density* of material in the Jupiter zone of the solar system. When Jupiter formed, it swept up all of this material. If one repeats this thought experiment for Saturn, Uranus, and Neptune, one finds that the surface density of the solar system decreases fairly smoothly out to Neptune, but beyond Neptune, with only small Pluto at 40 AU, the surface density crashes essentially to zero, like the edge of a compact disk. But real physical systems do not have sudden cutoffs unless such barriers are externally imposed on the systems. For example, the depth of a pond decreases steadily toward the shore, the density of the atmosphere decreases slowly upward into outer space, the heat felt from a campfire decreases slowly with distance from the flames, and the height of a pile of sand decreases slowly from peak to edge. We could create sharp edges to these structures: we could replace the pond with a swimming pool, take atmospheric gases and put them in a hot air balloon, encase our fire within the walls of a furnace, and put the sand in a sandbox. Then the water would go instantly from deep to zero depth, the air would go from dense to rarefied, the temperature from burning hot to cold, the sand depth from deep to zero. The walls are unnatural barriers that prevent the water, air, heat, and sand from doing what they would naturally do. So why does the density of matter in our solar system effectively end so suddenly, so unnaturally?

151

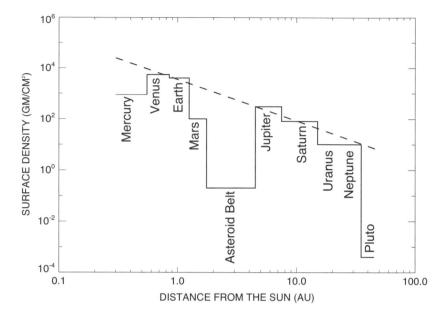

Figure 11.1: Plot of the surface density (grams per square centimeter) of the solar system at the time of formation, based on the amount of mass in each planet and the expected amount of hydrogen and helium gas that would accompany the solid particles bound into each planet. Effectively, we take the mass of each planet and smear it out uniformly around its orbital path around the Sun and then add to that mass the appropriate amount of gas. The asteroid belt is notice-ably deficient in mass, probably due to the influence of nearby Jupiter. Note how the surface density suddenly drops by at least one hundred times from Neptune to Pluto. Such a sudden drop is unphysical and suggests that a great deal of mass was in this region and is not accounted for in Pluto.

Either the solar system has a wall (if so, what is it?) or the surface density of matter in our solar system does not end abruptly. Kuiper decided that the solar system should not have a sharp outer edge and so hypothesized the existence of a disk of unseen (and therefore small and low-mass) trans-Neptunian objects, extending beyond 50 AU from the Sun.

Kuiper's idea of a disk whose surface mass density slowly de-creased all the way out to a few hundred AU was consistent with ideas first developed by Immanuel Kant and Marquis de Laplace in the eighteenth century that the solar system formed from an interstellar cloud that collapsed. This rotating cloud, collapsing under the influence of gravity, formed a star at the center and a flat disk orbiting that central star. The planets formed within this disk. While substantially improved over the two centuries since Laplace made this suggestion in 1796, our modern theory for how stars and planets form essentially follows this same idea. Thus, the young Sun presumably had a disk of gas and dust orbiting it. The disk would have been densest closest to the center and would have decreased in density slowly with increasing distance from the Sun. At some great distance, probably a few hundred AU from the Sun, the disk density would have decreased, effectively but not suddenly, to zero.

Edgeworth's and Kuiper's work was largely ignored and then forgotten. But good ideas tend not to disappear forever. In 1962, Alastair Cameron revived the idea that the outskirts of the solar system must be home to a great deal of material, left over from the epoch during which the planets formed.[3] Later that decade, Fred Whipple concluded that while a trans-Neptunian comet belt ought to exist, those comets were essentially undetectable. If the total mass of the objects in such a comet belt were large enough, the combined gravitational effect of all those comets would affect the orbits of Uranus and Neptune. Whipple concluded that such a belt also ought to influence the orbital behavior of comets, yet no such effects had been seen. In the 1970s, however, those effects would be uncovered.

In the 1940s, A.J.J. van Woerkom developed a theory for the orbital diffusion of comets due to the gravitational tugs of the giant planets. With this theory in hand, another Dutch astronomer, Jan Oort, recognized that many comets apparently came from vast distances (tens of thousands of AU) from the Sun. We recognize two types of comets, long-period comets and short-period comets. Long-period comets are seen once in human history; they have orbital periods of thousands of years or longer and therefore have semi-major axes of many hundreds, thousands, or even tens of thousands of AU. These enormous orbits explain why, to astronomers in the seventeenth century, most comets appeared to move in parabolic arcs. Short-period comets are those like Comet Halley, objects that have orbits small enough and orbital periods short enough to make possible multiple returns to the inner solar system in recorded history or even in a single human lifetime. A typical short-period comet has an orbital period of five to ten years (Comet Halley is unusual for a short-period comet, having a period of about seventy-five years).

Oort concluded that the Sun was surrounded by a vast, spherical cloud of comets that we now call the Oort comet cloud (although the idea was first suggested by Ernst J. Öpik in 1932). These comets orbit at distances of tens of thousands of AU from the Sun, in fairly circular orbits, preserved in the freezer of cold space far from the Sun. The outer edge of the Oort Cloud may extend as far as 50,000 AU from the Sun, nearly halfway to the nearest star. Over long periods of time, the gravitational influences of passing stars kick a few comets out of their stable orbits. Some of those new orbits will bring comets into the inner solar system where we see them. After an Oort Cloud comet is kicked inward toward the Sun and passes into the inner solar system, it will warm up. Surface ices will sublimate—that is, the ices pass directly from solids into gases—and the comet will develop the distinctive coma and tail appearance we associate with them. Then, following their highly elliptical orbits, many of these long-period comets will return to the Oort Cloud, where they hide for thousands of years, in frozen hibernation at the distant edge of the solar system while others will pass right through the Oort Cloud and into interstellar space, escaping the Sun's gravity completely and forever.

Short-period comets have orbital properties that are different from those of long-period comets. With orbital periods comparable to the orbital period of Jupiter, short-period comets not only must cross the orbits of the inner planets but also must have close encounters with Jupiter. With each passage into the inner solar system, a comet's icy surface heats up and the ices sublimate into a coma and tail. The material that enters the coma or tail is lost by the comet. As a comet loses mass, its orbit changes. In addition, after enough orbits, unavoidable close gravitational encounters with Jupiter and the inner planets doom such objects to short lives as short-period comets. After a few tens of thousands of years, such an object has

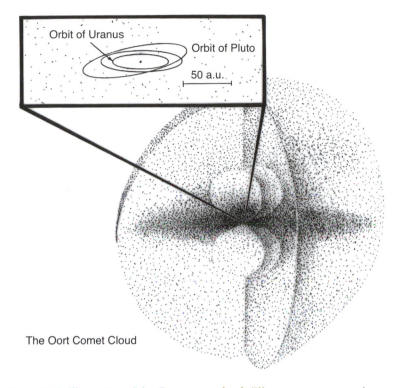

Figure 11.2: Illustration of the Oort comet cloud. (Illustration courtesy of JPL-Caltech)

shriveled to a tiny remnant of its original size, crashed into one of the planets or the Sun, or been slingshot by Jupiter out of the solar system. Since short-period comets have short lives, like long-period comets they must have a home, a freezer compartment within the bounds of the solar system from which they emerge. And where is that freezer?

In contrast to long-period comets, almost all short-period comets have prograde orbits (Comet Halley is the most notable exception) with low inclinations to the ecliptic plane. Edgar Everhart, in 1972, suggested that these orbital properties of short-period comets demand that they do not come from the Oort Cloud. They had to come from a reservoir confined to a volume of space near the plane

155

of the planetary system. Everhart showed that it was possible for a comet to be captured into the inner solar system from the Oort Cloud if those comets experienced gravitational encounters first with Neptune; in a stepwise process, a comet would be transferred inward, first by the gravitational influence of Neptune, which has a low inclination orbit, then of Uranus, Saturn, and finally Jupiter. Since the first step would occur near Neptune, most of these comets appear to come from low inclination orbits. Everhart's hypothesized mechanism did a good job of explaining the distribution of orbits of short-period comets, but it was remarkably ineffective: only one of every few thousand comets whose orbits might be changed by Neptune had a chance of making all the successful transfers into the inner solar system. Given the large number of short-period comets, if each one represented only a fraction of a percent of those perturbed from the short-period comet reservoir, and if objects have been becoming comets for the entire age of the solar system, there must be the equivalent of several Jupiter masses of comets in the Oort Cloud.

The trouble with Everhart's solution is that such a large mass in the Oort Cloud was untenable. No theories for forming the Sun and planetary system could possibly produce such a massive cloud of comets located tens of thousands of AU from the Sun. Everhart's solution was clever but wrong. What was needed was an even better idea, and that came soon thereafter from Julio Fernandez, of Uruguay. Fernandez took the Edgeworth and Kuiper idea and applied it to the Everhart model. Rather than assume that Neptune captures comets directly from the Oort Cloud, Fernandez suggested that Neptune captures comets from a closer comet reservoir located in the ecliptic plane in the trans-Neptunian zone. With the work of Fernandez, the idea of what became known as the Kuiper Belt was reborn;[4] further work in the late 1980s not only confirmed Fernandez's hypothesis but also led Martin Duncan and his colleagues to suggest that Chiron was a Kuiper Belt object in the process of diffusing inward, being passed from Uranus to Saturn and soon to Jupiter, at which time it will assume the look of an ordinary short-period comet (though perhaps a very big one).

THE SECRETS OF THE OUTER
SOLAR SYSTEM REVEALED

If Chiron is an object recently ejected from the Kuiper Belt, more such objects must exist, and soon they would be found. The Spacewatch project, in Arizona, was designed to find fast-moving, inner solar system objects, especially ones that might be on orbits near that of the Earth. On the ninth of January 1992, David Rabinowitz of Spacewatch discovered a very slowly moving object, one that clearly was an outer solar system object. First designated as 1992 AD, this object soon was named (5145) Pholus, Chiron's mythological brother. Like Chiron, Pholus is fairly large, with a diameter of about 200 km. Also like Chiron, Pholus has a very eccentric and unstable orbit. With a semi-major axis of 20.3 AU and an eccentricity of 0.57, Pholus's closest approach to the Sun is at a distance of only 8.7 AU, inside of Saturn's orbit; at its most distant orbital point, Pholus crosses outside of the orbits of Uranus and Neptune, reaching a distance of 31.9 AU. Thus, Pholus is a Saturn crosser, a Uranus crosser, and a Neptune crosser and is in a dramatically unstable orbit. Now, in addition to planets, moons, asteroids, and comets, we can add to the census of solar system objects the centaurs, of which there were now two known. Furthermore, we know that the centaurs represent objects scattered gravitationally inward from the Kuiper Belt; so, even though no Kuiper Belt objects had yet been found within the confines of the belt, the presence of two centaurs was a strong hint that another population of objects, Kuiper Belt objects, or KBOs, must exist but had simply escaped detection thus far.

Years before the discovery of Pholus, astronomer David Jewitt had been convinced that KBOs not only must exist but also could be found. He first began a survey, looking for KBOs, in 1986. For several years he found nothing. He perhaps was crazy to try. He had access to state-of-the-art detector systems on some of the biggest telescopes in the world, on Mauna Kea, Hawaii, but he didn't know where to look. He didn't know how big or bright a typical KBO object might be. And there were plenty of other projects that some-one with access to the same telescopes and equipment might have chosen to do, yet he persevered.

DISCOVERY OF THE KUIPER BELT

On August 30, 1992, working with Jane Luu, Jewitt was finally re-warded when he and Luu discovered 1992 QB$_1$ (this designation indicates that this object was the twenty-seventh asteroidal object to be reported (letter B with the subscript 1) during the second half of August (letter Q) in the year 1992. 1992 QB$_1$ was the first KBO known to be within the expected confines of the Kuiper Belt. The initial computations by Brian Marsden placed 1992 QB$_1$ at the in-credible distance of 37 to 59 AU from the Sun. Within a year, the orbit of 1992 QB$_1$ had been refined to give it an orbital semi-major axis (a) of 43.8 AU and an eccentricity (e) of 0.09; by 1995, the orbit was well understood: 1992 QB$_1$ has a semi-major axis of 44.0 AU, a fairly low eccentricity of 0.07, and a diameter of 280 km. The first KBO known to be inside the Kuiper Belt had indeed been found.

On March 28, 1993, six months after the initial discovery of 1992 QB$_1$, Jewitt and Luu found another Kuiper Belt candidate, 1993 FW. As with the discovery of 1992 QB$_1$, the first announce-ment of March 29 placed it between 38 and 56 AU from the Sun, with a crude estimate of a = 42.5 AU, and advised, "Similarity in motion and brightness to 1992 QB$_1$ suggests that 1993 FW is another Kuiper Belt candidate."[5] By early September of 1993, the orbital parameters were approximated as a = 43.9 and e = 0.04, which were very close to the best current values, a = 43.6, e = 0.05. The distance, in combination with the brightness and assumed density of 1993 FW, gave the object a diameter of 290 km. The Kuiper Belt now had a known population of two objects.

Another six months later, on two consecutive days, September 14 and 15, 1993, Jewitt and Luu rounded up two more Kuiper Belt candidates, 1993 RO and 1993 RP. The discovery announcements of September 18 and September 20 offered guesses at circular orbits for each object, with 1993 RO at a distance of 32.3 AU and 1993 RP at a distance of 35.4 AU from the Sun. By May 1994, astronomers assumed that, like Pluto, both were in orbits controlled and protected by the gravitational influence of Neptune, which led to the guess that both objects had semi-major axes close to 39 AU. Thus, 1993 RO was assigned a = 39.3 AU and e = 0.2 AU, very close to now-established values of a = 39.4 AU and e = 0.20, while 1993 RP was assigned values of a = 39.3 and e = 0.11. For 1993 RO, the May 1994 orbital parameters are identical to currently accepted values, giving 1993 RO a diameter of 140 km, given its measured brightness and assumed density. 1993 RP, however, has not been recovered since the discovery observations, so its orbit remains only a poor guess.

Only days after Jewitt and Luu's September 1993 discoveries, the observing team of Iwan Williams, Alan Fitzsimmons, and Donal O'Ceallaigh discovered 1993 SB and 1993 SC. The assumed, but innately indeterminate, orbits for these objects placed them in circular orbits 33.1 AU and 34.5 AU from the Sun. As for 1993 RO and 1993 RP, by May 1994, the orbits appeared better fit with semi-major axes close to 39 AU (for 1993 SC, the best-guess orbital parameters were a = 39.4 AU and e = 0.13, whereas for 1993 SB they were a = 39.3 AU and e = 0.18). By 1995, those orbits had been refined to a = 39.7 and e = 0.17 for 1993 SC, giving it an estimated diameter of 320 km, and a = 39.4 and e = 0.3 for 1993 SB, giving it an estimated diameter of 190 km.

In the span of twelve months, Jewitt and Luu had populated the hypothesized Kuiper Belt with four objects, Williams, Fitzsimmons, and O'Ceallaigh had added two more, and an entirely new part of the solar system had been discovered.

The work of Jewitt and Luu created a cottage industry of astronomical teams dedicated to finding trans-Neptunian objects (TNOs). Most of the TNOs are "classical KBOs," which are found between 42 and 48 AU from the Sun and have low eccentricities that help them maintain large separations from Neptune. Others are scattered disk objects (SDOs), which are found well outside of the region of the classical Kuiper Belt. Some of these objects have peri-

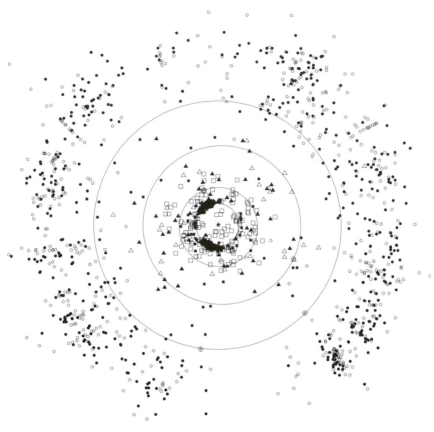

Figure 11.3: Plot showing the orbits of all known objects in the outer solar system (as of December 17, 2004). From the center outward, the orbits of the planets Jupiter, Saturn, Uranus, and Neptune are shown as circles and the current location of each planet is marked by the symbol of a plus sign enclosed in a circle. Pluto is marked by the planet symbol just off the Neptune circle at the bottom of the plot. The current locations of the minor bodies of the outer solar system are shown with different symbols to denote different classes of objects. Centaurs and unusual, high-eccentricity objects are shown as triangles; Plutinos, scattered disk objects, and "classical" or "main belt" Kuiper Belt objects are shown as small circles; numbered periodic comets are shown as squares. Objects observed at more than one opposition (i.e., observations separated by at least one year) are shown by filled symbols; objects observed at only one opposition (whose orbits are less well known) are shown by open symbols. (Illustration used by permission of Gareth Williams and the Minor Planet Center)

helia inside of 30 AU and thus cross Neptune's orbit but are not protected by Neptune, while others have aphelia far beyond 50 AU; some appear to be in the process of being removed from, or scattered out of, the Kuiper Belt. As of December 2005, the numbers of such objects had proliferated, with 59 centaurs, 100 SDOs, and 925 other TNOs known.[6] At the current pace of discovery, more than 100 centaurs, SDOs, and TNOs are discovered every year, though some of them (as many as 40 percent) are lost when they are not recovered in follow-up observations.

THE GIANTS OF THE KUIPER BELT

The biggest such objects among these newly discovered outer solar system objects include Varuna (also known as minor planet number 20000; originally designated 2000 WR_{106}), Ixion (#28978, or 2001 KX_{76}), 2002 AW_{197} (#55565), Quaoar (#50000, or 2002 LM_{60}), 2002 TC_{302} (#84522), 2003 EL_{61}, 2003 UB_{313}, Sedna (#90377, or 2003 VB_{12}), Orcus (#90482, or 2004 DW), and 2005 FY_{9}.

Varuna, a classical KBO discovered on November 28, 2000, by Robert S. McMillan using the Spacewatch Telescope in Arizona, has a semi-major axis of 43.3 AU, a nearly round orbit with an eccentricity of only 0.05, and a diameter estimated in the range of 760 to 1,040 km, with a best estimate of 900 km.

Ixion, discovered with a telescope in Chile on May 22, 2001, by a team led by Robert Millis of Lowell Observatory, has a Pluto-like semi-major axis of 39.5 AU and an elongated orbit with an eccentricity of 0.24, and is estimated to have a diameter between 900 and 1,230 km, with a best estimate of 1,065 km.

2002 AW_{197} (the large number in the subscript reveals the rapid rate of discovery of small objects in the solar system; object 2002 AW_{197} was the 4,947th object to receive a Minor Planet Center designation in the first half of January 2002), another classical KBO discovered by Michael Brown and Chadwick Trujillo of Caltech, has a semi-major axis of 47.5 AU and an eccentricity of 0.13; thus, its orbit stretches from 41 to 53 AU. J. L. Margot, working with Brown

Figure 11.4: Image of the Kuiper Belt object Quaoar, at a distance of 42 AU from the Sun. (Image courtesy of NASA and M. Brown, Caltech)

and Trujillo, estimated the diameter of this object between 770 and 1,010 km, with a best estimate of 890 km. John Stansberry and his collaborators, however, used observations made in 2004 with the Spitzer Space Telescope to downsize the estimated diameter of 2002 AW_{197} to 700 km. The smaller size estimate, made by Stansberry, is a consequence of measurements that indicate 2002 AW_{197} is better at reflecting sunlight than are comets (comets typically reflect about 4 percent of the sunlight that hits their surfaces; 2002 AW_{197} apparently reflects 18 percent of the incident sunlight).

Quaoar, another classical KBO discovered on June 4, 2002, also by Brown and Trujillo, is the first of the giant Kuiper Belt objects whose size has been measured rather than inferred. Quaoar has a semi-major axis of 43.4 AU and an eccentricity of only 0.03. Brown and Trujillo were able to use the Advanced Camera for Surveys on the Hubble Space Telescope to measure the diameter of Quaoar—forty-thousandths of one second of arc. At its current distance of about 42 AU, this angular size corresponds to a physical size of over 1,300 km.

2002 TC_{302} was discovered on October 9, 2002, at Palomar Observatory by the team of Raymond Bambery, Eleanor Helin, Steven Pravdo, Kenneth Lawrence, Michael Hicks, and Robert Thicksten. With a semi-major axis of 55.1 AU and an eccentricity of 0.29, it has an estimated size of about 1,200 km.

2003 EL$_{61}$ was discovered on March 7, 2003, with the Telescopio Sierra Nevada in Granada, Spain, by F. J. Aceituno, P. Santos-Sanz, and J. L. Ortiz. They announced their discovery on July 28, 2005. Given the distance (2003 EL$_{61}$ has a semi-major axis of 43.3 AU and an eccentricity of 0.19, making it a classical KBO with an orbital period of 297 years) and brightness (about ten times fainter than Pluto) of 2003 EL$_{61}$, the Spanish team estimated the size of 2003 EL$_{61}$ as comparable to that of Sedna, approximately 1,500 km in diameter. On January 26, 2005, using one of the two 10 m telescopes at the Keck Observatory in Hawaii, Brown, Trujillo, and Rabinowitz discovered that 2003 EL$_{61}$ has a moon; through follow-up observations over the next five months, they determined that the moon was in a forty-nine-day orbit. From the size and period of the orbit and the relative brightness of the moon and 2003 EL$_{61}$, Brown's team determined that 2003 EL$_{61}$ has a mass nearly 32 percent that of Pluto and a diameter close to 70 percent that of Pluto (1,500 km), in agreement with the size estimate of Aceituno, Santos-Sanz, and Ortiz.

2003 UB$_{313}$, an SDO, was discovered by Michael Brown on January 5, 2005, in data obtained on October 21, 2003, by Brown, Trujillo, and Rabinowitz, using a forty-eight-inch telescope at Palomar Observatory. As of the time of the announcement of the discovery of 2003 UB$_{313}$, on July 29, 2005, 2003 UB$_{313}$ was the most distant known object ever seen in orbit around the Sun, at a present-day distance of 97.5 AU, which also appears to be the most distant point in its orbit from the Sun. The 560-year orbit of 2003 UB$_{313}$ (semi-major axis of 67.7 AU, eccentricity of 0.44) will take it from 97 AU inward to only 38 AU. On the basis of the brightness and distance of 2003 UB$_{313}$, Brown, Trujillo, and Rabinowitz have suggested that 2003 UB$_{313}$ is the largest known object in the solar system beyond Neptune, with a diameter 25 percent or more greater than that of Pluto (2,700 km)! Hence, if Pluto is a planet, then 2003 UB$_{313}$ also is a planet and October 21, 2003, marks the day when the (most recent) tenth planet in our solar system was found.

Sedna is an SDO, discovered on November 14, 2003, by Brown, Trujillo, and Rabinowitz, of Yale University, at Palomar Observatory.

Figure 11.5: Image of the Kuiper Belt object Sedna, at a distance of 86 AU from the Sun. (Image courtesy of NASA, ESA, and M. Brown, Caltech)

Current estimates indicate that Sedna has a semi-major axis of just over 500 AU, an eccentricity of 0.85, and a diameter of roughly 1,500 km. Consequently, Sedna comes as close as 75 AU to the Sun and travels out beyond 900 AU and requires 10,500 years to orbit the Sun!

Orcus was discovered by Brown, Trujillo, and Rabinowitz at Palomar Observatory on February 17, 2004. Orcus has a semi-major axis of 39.9 AU and an eccentricity of 0.27, giving it a Pluto-like orbit. Brown, Trujillo, and Rabinowitz's initial estimate indicates that Orcus could be as much as 10 percent larger than Quaoar and comparable in size to Sedna.

2005 FY$_9$, was discovered by Brown, Trujillo, and Rabinowitz on March 31, 2005. It has a semi-major axis of 45.7 AU, an eccentricity of 0.16, and an orbital period of 308 years, making it a classical KBO. Given that it is a tiny bit brighter than 2003 EL$_{61}$, it is almost certainly of comparable size to, or perhaps a bit bigger than, 2003 EL$_{61}$.

Figure 11.6: Comparison of sizes of Sedna (diameter estimated as up to 1,500 km), Quaoar (diameter estimated as up to 1,300 km), Pluto (diameter = 2,300 km), the Moon (diameter = 2,172 km), and Earth (diameter = 12,756 km). (Illustration courtesy of NASA/JPL-Caltech and R. Hurt, SSC-Caltech)

A number of other outer solar system objects may also have sizes in the 600–1,000 km range, similar to that of 2002 AW_{197}. These include, among others, 1995 SM_{55} (#24835), Chaos (#19521, or 1998 WH_{24}), 1999 DE_9 (#26375), 1999 TC_{36} (#47171), 1999 TO_{66} (#19308), Huya (#38628, or 2000 EB_{173}), 2002 VE_{95} (#55638), 2002 UX_{25} (#55637), 2002 TX_{300} (#55636), 2003 AZ_{84}, 2003 VS_2 (#84922), 2004 GV_9 (#90568), and 2004 XR_{190}.

With the rapid discovery, beginning with that of 1992 QB_1, of hundreds of objects in the trans-Neptunian region, the Kuiper Belt has become a known part of the solar system and a more distant, colder analog of the asteroid belt. Varuna, Ixion, 2002 AW_{197}, Quaoar, 2002 TC_{302}, 2003 EL_{61}, 2003 UB_{313}, Sedna, Orcus, and 2005 FY_9 are all comparable in size to, and several are almost certainly as much as 50 percent larger in diameter than, Ceres, which has a diameter of about 925 km. At least one, 2003 UB$_{313}$, is likely larger than Pluto.

Post facto, early nineteenth-century astronomers realized that Ceres, the largest of the asteroids, was fortuitously the first-discovered member of a whole population of objects that collectively make up the asteroid belt. Historically, the discovery of the largest KBOs known thus far postdates the discovery of the Kuiper Belt itself. The first KBOs discovered were small, only a few hundred kilometers in diameter, too small for anyone to consider them anything but big cometlike objects. Varuna, Ixion, and Sedna fit naturally into the population of KBOs as the largest known among this family of objects. If Ixion had been discovered first, perhaps in 1979, would it have been identified as the tenth planet? Like Ceres, would it have then lost this exalted status with the subsequent discovery of Varuna and the other KBOs? Does 2003 UB$_{313}$ merit being labeled the tenth planet?

12
The Plutinos

O

Like Pluto, Ixion crosses Neptune's orbit, reaching a perihelion distance of 29.75 AU, and travels as far out as 48.94 AU, comparable to Pluto's perihelion and aphelion distances of 29.69 and 49.37 AU. In fact, dynamically, these two objects are difficult to distinguish. Ixion has a semi-major axis of 39.34 AU and a period of 246.77 years, whereas Pluto has a semi-major axis of 39.53 AU and a period of 248.54 years. The principal difference between the two objects would appear to be size, with Ixion having a radius of about half that of Pluto.

A second object in the outer solar system, Charon, also shares similar characteristics with Pluto and Ixion. Charon, though a moon of Pluto, orbits the Sun, albeit while also orbiting Pluto, and thus has the same orbital parameters as Pluto. Charon is smaller than Pluto, with an estimated diameter of 1,186 km. Charon, in other words, is essentially indistinguishable from Ixion. However, we label one as a Kuiper Belt object while the other is designated a moon of Pluto.

It turns out that Ixion is not the only KBO with an orbit extremely similar to that of Pluto and Charon. Objects in this special subgroup of KBOs are collectively known as Plutinos. Although the exact orbital boundaries for Plutinos are not firmly defined, Plutinos have orbits with semi-major axes between approximately 39 and 40 AU, with 95 percent of these having semi-major axes between 39.1 and 39.8 AU. Inside of 39 AU and extending inward to Neptune,

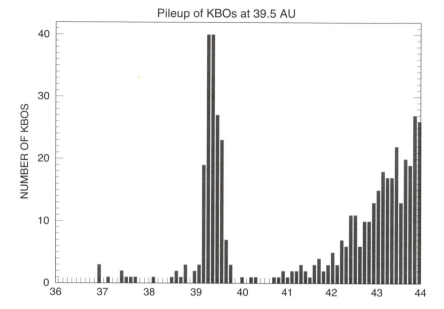

Figure 12.1: Most of the volume of the solar system in between Neptune and the Kuiper Belt is empty of orbiting objects, except for the Plutinos, which have orbital periods and semi-major axes nearly identical to those of Pluto (at 39.5 AU).

there are very few KBOs. Beyond 40 AU, there are almost no known KBOs out to 41 AU; farther out, the number of KBOs starts to climb. Here, beginning at distances of about 42 AU, is where the main Kuiper Belt objects orbit. The Plutinos appear anomalous.

On the basis of Kepler's third law (recall that the square of the orbital period, measured in units of years, is equal to the cube of the semi-major axis, measured in units of AU), we know that all the Plutinos must have orbital periods almost identical to that of Pluto, since they have semi-major axes nearly identical to that of Pluto. In addition, most, but not all, have moderately large orbital eccentricities, and so, like Pluto, their orbits cross the orbital path of Neptune or have the potential to have close encounters with Neptune.

The centaurs, those objects that have been tossed inward from their original locations in the Kuiper Belt into smaller orbits, are on planet-crossing orbits and therefore must have short lifetimes; they are destined to have close encounters with one or more giant planets that will slingshot them into even more unstable orbits within, or eject them entirely from, the solar system. Short-period comets are on planet-crossing orbits or come close enough to Jupiter that they also have short dynamical lifetimes. Asteroids known as Earth-crossing asteroids are destined to collide with the Earth or Moon or to be tossed into the Sun or into another unstable orbit. Perhaps Pluto and all the other Neptune crossers and near crossers eventually must collide with Neptune or experience a close encounter with Neptune that will toss them out of a Neptune-crossing orbit.

In fact, this will *not* happen to the Plutinos.

PROTECTING THE PLUTINOS

The special condition that prevents Neptune-Pluto and Neptune-Plutino collisions tells us a great deal about the history and contents of the solar system. Pluto and the Plutinos are protected by a circumstance known as an orbital resonance. Depending on the exact details of the situation, an orbital resonance will either stabilize or destabilize an orbit. In the Neptune-Pluto and Neptune-Plutino case, the resonance has saved Pluto and the Plutinos from doomsday collisions with Neptune.

The orbital periods of Pluto (248.54 years) and Neptune (164.82 years) are very nearly in the ratio of three to two (3:2). All the Plutinos also share this relationship, called the 3:2 orbital resonance, with Neptune and thus all of these objects complete almost exactly two orbits around the Sun for every three orbits of Neptune. What does this special situation mean for Neptune and Pluto (and the Plutinos)?

Imagine Neptune and Pluto are runners at the starting line and in the inner and outer lanes, respectively, of a two-lane, oval racetrack. Neptune is monstrously large, seven feet five inches tall, weighing 375 pounds, all of which is muscle; Pluto is petite, four feet seven inches tall, barely 85 pounds, several of those pounds being her waist-length hair. As the starting gun for the race is fired, Neptune rushes into the lead, setting a pace that is one and a half times faster than the speed Pluto can muster. Steadily, Neptune opens up more and more space between himself and trailing Pluto. By the time Pluto completes a single lap around the track, Neptune has already completed one and one-half laps and is on the opposite side of the track, with a lead over Pluto of a half lap. Now, steadily, Neptune starts to gain on Pluto. After completing three laps to Pluto's two, Neptune catches up with and laps Pluto. Given the steady pace of each racer, the next time Pluto and Neptune are close together on the track will be when Neptune next laps Pluto, after Neptune completes three more laps around the track. In fact, the *only* place on the track when Neptune and Pluto will ever be near each other is at the starting line.

Now let's change this scenario a little. Instead of two separate lanes on the track, we build a track in which the two lanes are separated by a median strip around most of the track, forcing both runners to stay in their respective lanes. However, at a single spot on the track, the two lanes merge into one and pass through a small tunnel, similar to the crossover section of a skating oval on which Olympic long-track speed skaters race. If the merge point, the tunnel, is at the starting line, then when the starting gun sounds, Neptune and Pluto will burst forth into their own lanes until Pluto completes two laps and Neptune completes three. As Neptune pulls closer to Pluto near the end of Neptune's third lap and Pluto's second, the two runners merge into a single lane and into the tunnel; just as Neptune catches up with Pluto, the two lanes converge inside the narrow tunnel. Neither runner is willing to slow down and yield the lead to the other through the single-lane tunnel. Inside the tunnel, the runners collide. Their legs tangle and, when 375-pound Neptune collides with wispy Pluto, Neptune tramples Pluto, crushing her into the wall of the tunnel. Pluto collapses; she's out of the race. Only Neptune, unscathed, emerges from the tunnel.

This arrangement, two running lanes but a single crossover point, is almost never well suited for the survival of the smaller runner (or the smaller planet); however, we can reengineer the design of the track so that both runners can run forever, provided the runners agree to run at the perfect speeds. Let's move the single-lane tunnel to a position exactly halfway around the track from the starting line. Now, when Neptune enters the tunnel on the first lap, still running one and one-half times faster than Pluto, Pluto is one-sixth (sixty degrees) of the circumference of the track behind Neptune, so no collision occurs. The next time Neptune enters the tunnel, having run exactly one and one-half laps, Pluto has run only one lap and has just reached the starting line. No collision occurs. The third time Neptune enters the tunnel, having run two and one-half laps, Pluto has run one and two-thirds laps and is one-sixth of the circumference of the track ahead of Neptune, about to be lapped again. Finally, Neptune catches up with and safely passes Pluto at the starting line. If they keep running, they will never be near each other in the tunnel. The 3:2 ratio of their speeds keeps them apart, never closer than an angular distance of sixty degrees when one of them enters the tunnel, and protects them from a collision that would knock Pluto out of the race.

This is exactly the situation with Pluto and Neptune. When Pluto approaches perihelion (the tunnel position), crossing paths with Neptune and becoming the "eighth" planet for two decades out of every 250 years, Neptune is always at least sixty degrees away from Pluto when Pluto approaches perihelion. As long as the ratio of the orbital periods of Pluto and Neptune stays very close to 3:2, and as long as the crossover point is far from the starting line, the two planets will never collide.

On the other hand, what if we had a trans-Neptunian object (TNO), a Plutino wanna-be, whose orbital period was nearly exactly one and one-half times greater than that of Neptune, but the crossover point for this TNO was at a position at which Neptune and the TNO were fairly close together? Then Neptune and the TNO will come close enough together that their gravitational attraction for each other will affect their orbits. Neptune will pull the TNO forward, giving it more energy; with more energy, the orbit of the TNO will change. The semi-major axis will increase, as will the orbital period. The next time Neptune and this TNO lap each other, they will be closer to the crossover point, their gravitational influence on each other will be greater still, and their orbits will be affected again. After a few tens or hundreds or thousands of orbits, perhaps after only a few million years, these two objects will arrive at the crossover point together and collide, or they will come so close together that Neptune will slingshot the TNO out of the solar system or into the Sun. No matter the exact details of the final event, any TNO whose crossover point was too close to Neptune would have found itself ejected from the Plutino zone early in the history of the solar system. Such a TNO would have started with some but not all of the right conditions for a Plutino and thus could never become a Plutino.

What about a Plutino wanna-be for which the ratio of orbital periods with Neptune was 1.465 (2.93:2), very close to but not quite 1.50 (3.00:2), but for which the crossover point is perfectly placed halfway around the track from the starting point? Since the orbital ratio is not exactly 3:2, after 2 laps of this TNO, Neptune has completed only 2.93 laps. Rather than lap the TNO at the starting line, Neptune finally catches up with the TNO about 54 degrees past the starting line. Just over 3 laps later, Neptune laps the TNO again, this time 108 degrees past the starting line. Steadily, the place where the lapping event occurs moves around the track, to 162, 216, and 270 degrees past the starting line after 15 laps. Eventually, after 30 laps, the crossover point migrates into the tunnel and a collision occurs. Consequently, unless the ratio of the orbital period of a TNO to that of Neptune is extremely close to 1.50, the TNO will be tossed out of the Plutino zone, fated never to become a Plutino.

Almost no TNOs have orbital periods between 225 and 243 years or 253 and 272 years. This is the Plutino desert, cleansed of Plutino wanna-bes. What are the odds that when the solar system formed, all the objects with orbital semi-major axes near 39.5 AU happened to be Plutinos and that no objects ever existed at 38.5 or 40.5 AU? Zero. In comparison, what is the likelihood that many other objects formed early in the history of the solar system but only the Plutinos survived? High. The Plutinos are the survivors. If they are the survivors, then four and a half billion years ago, this part of the solar system must have been full of many balls of ice, hundreds or thousands of kilometers in diameter. Amazingly, additional evidence in the outer solar system lends support to this idea, first proposed by American astronomers Alan Harris, at a conference celebrating the fiftieth anniversary of the discovery of Pluto, in 1980, and by Alan Stern, in 1990, that the region of the solar system that encompasses Pluto was once home to thousands of Pluto-like objects. Triton and Charon may provide that additional evidence.

THE TROUBLE WITH TRITON

Triton is the largest moon of Neptune. Given that the orbits of Pluto and Neptune cross, at many times Triton and Pluto are at similar distances from the Sun. Triton has a diameter of 2,700 km. The most similar objects in the solar system to Triton, based on size alone, are Pluto (diameter 2,274 km, barely 10 percent smaller) and Europa (diameter 3,138 km, 10 percent larger). Europa, one of the four large moons of Jupiter, however, is only 5 AU from the Sun and is composed of about equal parts rock and ice; in contrast, Triton and Pluto are almost entirely composed of ices. In fact, Triton and Pluto have densities nearly identical to each other, about twice that of water ice. Triton and Pluto share other characteristics: they show similar materials at their surfaces, these being nitrogen, carbon monoxide, methane, and water ice. As Alan Stern, of the Southwest Research Institute in Boulder, Colorado, has pointed out, "[T]he planet Pluto and satellite Triton are more similar to one another than are any two planets in the solar system."[1] Triton formed in the cold outer parts of the solar system, whereas Europa formed near Jupiter. By size and composition and place of origin, Triton and Pluto are siblings; Europa and Pluto are not.

Triton is a very unusual moon because its orbit is retrograde; that is, it orbits Neptune in the opposite direction from that in which the planets orbit the Sun and virtually all the moons in the solar system orbit their parent planets. Of the large moons in the solar system, only Triton has a retrograde orbit. To make sense of this, we have to ask why a retrograde orbit is unusual, and to do this we have to remember how astronomers believe planetary systems form.

Figure 12.2: Image of the surface of Neptune's largest satellite, Triton, obtained by *Voyager* 2. Image resolution is 900 meters. The imaged region, which is about 500 km across, shows old impact basins modified by flooding, melting, faulting, and collapse. Only a few impact craters dot the area, indicating that internally driven geological processes dominate the sculpting of Triton's surface. (Image courtesy of NASA/JPL-Caltech)

The critical issue to understand, in this context, is that systems of planets form in flattened disks of gas and dust around newborn stars. These flattened disks form because in a rotating interstellar cloud that is collapsing toward its center under the influence of gravity, most of the cloud material has too much rotational energy, too much angular momentum, to make it all the way to the center. Instead of collapsing from a large sphere to a smaller sphere, the spherical cloud pancakes, forming a central star surrounded by a disk. Because the entire system formed from a cloud in which all the cloud particles were revolving in a single direction around the center of the cloud, all the material in the disk that forms must be revolving around the central core, the newborn star, in the same direction. Thus, all the planets and asteroids and virtually all the comets revolve around the Sun in the same direction. For the same reason, any moon that formed in a disk around a newborn planet will revolve around the planet in the same direction as the planet rotates and the same direction in which the planet revolves around the Sun.

Given our understanding of how planets and moon systems formed, the only way for a moon to find itself in a retrograde orbit is for the planet and moon to form separately and for the planet to capture the moon into orbit during a close encounter. Imagine that Triton was once in a Pluto-like orbit, crossing Neptune's orbit like a Plutino wanna-be. But Triton's passage inside of 30 AU occurred so close to Neptune that it either passed through the upper layers of Neptune's atmosphere, thereby losing enough energy to be captured but not destroyed by Neptune, or collided with one of Neptune's moons, destroying the ill-fated moon but, again, losing enough energy to be captured into orbit around Neptune. Since the Triton-Neptune close encounter could easily occur with Triton slightly ahead of or behind Neptune in Neptune's orbit, Triton could be captured into either a prograde or retrograde orbit, depending on the exact conditions of the close encounter. The important thing is that no matter what the method of capture, Triton had to have been captured by Neptune from a solar orbit. Therefore, *Triton was once a planet*. What is the likelihood, Alan Stern suggested, that Neptune captured into a retrograde orbit one of the only two objects in the Pluto zone, Pluto being the other, with diameters greater than 2,000 km? Isn't it much more likely that there were thousands or tens of thousands of such objects, thereby making the capture of Triton as a moon of Neptune very likely rather than extremely unlikely?

WHAT ABOUT CHARON?

With a diameter that is almost exactly half that of Pluto, a distance from Pluto of only 19,600 km (barely eight times the diameter of Pluto), and a mass about one-eighth that of Pluto, the Pluto-Charon system is considered by some astronomers to be a binary planet.

The next closest planet-moon system to Pluto-Charon in these characteristics is the Earth and Moon. The Earth's moon, though it is about one-quarter of the diameter of the Earth, is only eighty-one times less massive than the Earth and orbits at a distance equal to 61 Earth radii. We can say colloquially that "the Moon orbits the Earth," because even though Isaac Newton taught us that the Moon and Earth both orbit a point that falls somewhere in between the centers of both objects, that point, the *barycenter* of the Earth-Moon system, lies slightly closer to the center of the Earth than the surface of the Earth; that is, the distance from the center of the Earth to the barycenter of the Earth-Moon system is less than the radius of the Earth. With the Pluto-Charon system, the barycenter is actually located in between the two objects. If we watched Pluto and Charon carefully, we would see both objects orbiting around an empty spot in space; that spot is closer to Pluto than to Charon but is well above the surface of Pluto. This criterion, that the center of the orbits of both the bigger and smaller objects in a double-object system lies somewhere in between the two objects, is a good reason to label Pluto-Charon a double-planet system rather than a planet-moon system.

Whereas we understand that the major moon systems of Jupiter, Saturn, Uranus, and Neptune (except for Triton) formed from material in disks around those planets, many details about Earth's Moon make such a formation scenario impossible. The most important of these details is the composition of the Moon. The Moon resembles the composition of the outer parts of the Earth, but does not resemble the average composition of the entire Earth. The Moon is deficient in iron and iron-loving elements (e.g., gold, platinum, iridium) compared to the large iron and siderophile content of the Earth and is also deficient in volatiles, these being elements and molecules (e.g., zinc, cadmium, tin, nitrogen, carbon, water) that vaporize at relatively low temperatures. In the Earth, most of the iron is in the core, having sunk to the center of the Earth during a phase very early in the history of the Earth when the entire planet was soft, such that most of the densest materials sunk under the influence of gravity. If the Moon formed from the same materials as the Earth, from a disk of material orbiting Earth, it should have the same overall composition as the Earth, including the iron. In 1976 and 1977, William Hartmann, Don Davis, and Alastair Cameron first proposed the giant impact hypothesis for the formation of the Moon that adequately explains the iron and volatile deficiencies in the Moon. According to this theory, after the Earth was nearly fully formed and the core had formed, a Mars-sized object collided with the Earth. Such collisions were a natural part of the formation process of the planets, but near the end of the planet formation epoch, the collisions were among fairly large objects. During this collision, the Mars-sized impactor was destroyed. Sophisticated simulations of such collisions done by Cameron, Willy Benz, and Jay Melosh in the late 1980s and early 1990s and by S. Ida, Robin Canup, Glen Stewart, E. Kokubo, J. Makino, H. Miura, and K. Ichinose beginning in the mid-1990s indicate that most of the material from the impactor, along with a fair amount of the proto-Earth, splashed out of the impact site. Much of the debris went off into orbit around the Sun, and some of it fell back to Earth; however, some of the debris found its way into orbit around the Earth. Quickly, this orbiting material coalesced to form a single object, our Moon. Since most of the debris from the collision was splashed from the outer parts of the young Earth, inside of which the iron had already settled to the core, the debris was iron poor, hence the Moon is iron poor. In addition, most of the debris would have been superheated, which would naturally lead to the loss of materials that vaporize at low temperatures.

Charon turns out to be less dense than Pluto. In fact, Charon appears to have a higher water content than Pluto and has a surface that appears to be almost entirely crystalline water. Like the Moon, the composition of Charon is unusual, and the explanation for the origin of Charon may be similar to that of the Moon. Charon may have formed when an impactor, perhaps with 10 percent the mass of Pluto, crashed into Pluto. The impactor was too small to destroy Pluto but large enough to splash a great deal of material off the surface of Pluto, lofting much of that material into orbit where it formed into Charon. If Pluto, like the Earth, had already had its heavier elements settle toward the core, the outer layers would have been dominantly water ice; thus, Charon would be enriched in water relative to Pluto.

Collisions in the inner solar system, especially in the early days of the solar system, happened often. The volume of space is small, and so the density of objects in the region is high; like cars on a crowded city freeway, collisions are likely and occur often. In contrast, in the outer solar system, the volume of space is so immense that the space density of objects is very low, like camels in the Sahara desert. Thus, collisions must have been extremely rare. Yet, the existence of Charon indicates that one such collision may well have occurred. Alan Stern calculated the odds of such a collision and found that the only way that Pluto might have been hit by an object that could have triggered the formation of Charon would be if between three hundred and three thousand objects capable of forming Charon were in or passing through the Pluto zone, several billion years ago, thereby increasing the odds that at least one such collision would occur. If one of these objects became Pluto, a second triggered the formation of Charon, and a third became Triton, where have all the others gone?

If we define the Plutinos as objects with orbital periods within about 2 percent that of Pluto (i.e., from 39.0 to 40.0 AU; this covers all objects near the 3:2 orbital resonance with Neptune), we find more than 150 objects. Almost certainly, all of these are objects with diameters of a few hundred to a thousand kilometers in orbits that are protected by the orbital resonance with Neptune. Equally likely, the Plutinos already found are only a fraction of those that still exist, and those that still exist are only a fraction of the Plutino wannabes that were in orbits that were unprotected from the gravitational tugs of Neptune, exerted steadily over hundreds of millions of years. Only the lucky survived.

Pluto, Charon, and Triton were lucky. So were 1999 HK_{151}, 2000 FV_{53}, 2001 UO_{18}, and all the other yet unnamed Plutinos. As best as we know it at this time, Pluto is the largest of the Plutinos in primary orbits around the Sun; the biggest Plutino, however, may be Triton.

Why, then, do we call Triton a moon and Pluto a planet?

13
Is Pluto a Planet?

○

Once upon a time, our ancestors believed that the Sun and Moon, along with Mercury, Venus, Mars, Jupiter, and Saturn, were planets—these being the *wandering stars that circled the Earth*—and that the Earth was a special object, fixed at the center of the universe, with the universe comprising the entirety of the known heavens. But they were wrong. Aristotle's aether does not exist, and celestial bodies do not move in perfect circles around the Earth, having achieved their telos.

In the sixteenth century, Copernicus reinvigorated an old idea, that the Earth orbits the Sun. For Copernicus, the known universe necessarily became much larger, as the stars had to be much more distant than had been thought, in order to explain the absence of stellar parallax. Soon thereafter, Tycho Brahe, Kepler, Galileo, and Newton provided observational and theoretical support that ultimately demoted the Earth from the central object in the universe into a mere planet; the change in the status of the Earth occurred in parallel with the promotion of the Sun into the object at the center of the solar system (and, as understood in the seventeenth century, the universe). As a result, in the late sixteenth and very early seventeenth centuries, we had a new way of defining planets— planets were now understood to be *the Earth plus all the wandering stars except the Sun*—and we had seven of them, although two of them—the Earth and Moon—were part of a double-planet system, and though not very many scholars of any variety accepted this new, heliocentric idea.

Kepler's monumental mathematical discoveries of 1609 provided a new way of thinking about the planet problem that distinguished between the Earth's moon and the other six Copernican planets. Planets were objects that obeyed Kepler's laws; that is, they orbit the Sun in elliptical paths with the Sun at one focus of their elliptical orbits.

Throughout the seventeenth century, the sensational observational discoveries made by Galileo, Cassini, and Huygens opened our eyes to the presence of objects we now call moons, around Jupiter and Saturn. For most of this century, all of these moving objects were considered planets; but by the end of the century, astronomers were distinguishing between objects orbiting the Sun and objects orbiting objects that orbit the Sun. By 1800, Earth's moon had metamorphosed from a planet into a moon. By doing so, the astronomers of the seventeenth century added a whole new class of objects to the solar system and reduced the number of planets to six. These first astronomers to be armed with telescopes provided, for the first time since Aristotle's definition failed, a clear definition of a planet: planets were *all of the objects whose primary orbits were around the Sun.*

Halley, quick on the heels of Newton, showed that planets shared the privilege of orbiting the Sun with more ephemeral objects called comets. Thus, by the mid-eighteenth century, astronomers recognized the existence of objects that orbited the Sun but that were most certainly not planets. Thus, by 1760, the definition of *planet* was muddled once again.

In 1781, by accident, Herschel discovered the planet Uranus, giving the solar system, once again, a seventh planet. Inspired by the Titius-Bode law, the Celestial Police made plans to find the presumed missing eighth planet, but in 1801, before his colleagues had contacted him and asked him to join the hunt, Piazzi found Ceres. Now the solar system had eight planets. Furthermore, Uranus and Ceres helped redefine *planet* in a meaningful and unique way: a planet is *a body that orbits the Sun according to Kepler's laws and obeys the Titius-Bode law.*

Quickly, European astronomers found additional objects like Ceres: in 1802 Pallas became the ninth object thought of as a planet, Juno (1804) the tenth, and Vesta (1807) the eleventh. Then slowly, over a half century, but largely triggered by the discovery in 1845 of Astraea, the twelfth object known as a planet, a consensus developed within the astronomy community that Ceres and the other newly discovered objects should not be called planets (the terms *minor planets* and *asteroids* were both used thereafter) and that an entire belt of Sun-orbiting asteroids existed. The big losers in this cycle of discovery were Ceres, demoted from the eighth known planet to the biggest of the asteroids, and Piazzi, who is much less famous for having discovered the first and largest asteroid than he would have been for discovering the eighth planet.

Forty-six years after the discovery of Ceres, careful work in mathematical astrophysics led to the discovery of Neptune, which was the thirteenth object known as a planet in the year of its discovery. But by the end of the 1840s, with the recognition of the existence of the asteroid belt, Neptune became known as the eighth planet, albeit the second object, after Ceres, to merit that designation. By 1850, the known solar system included a star, eight planets (again), at least fourteen moons, a large number of comets and asteroids, and no sensible definition of a planet (again).

Soon thereafter, overeager astronomers discovered, for the fourth time, a ninth planet, the trans-Neptunian planet, known only as Hind's planet, and, for the third time, a tenth planet, the intra-Mercurial Vulcan. As a result, the solar system had ten planets again, albeit for only two decades.

In 1879, Hind's planet was relegated to the dustbin of history. Again, the solar system had only nine planets.

By 1910, Vulcan too was gone, having been no more than the creation of the too-active imagination of an amateur astronomer and the too-ready-to-believe attitude of the esteemed Urbain Le Verrier. By 1915, Einstein had shown that the previously unexplained precession of the perihelion of Mercury resulted from the curvature of space-time near the Sun, not from the presence of an unseen planet. Again, the solar system had only eight planets.

In 1930, Clyde Tombaugh discovered Pluto, the fifth ninth planet. The director and astronomers at Lowell Observatory made sure, through carefully crafted public announcements, that the world knew that the long-searched-for Planet X had been found, that Percival Lowell's theoretical findings had been fulfilled. On March 13 and 14, 1930, newspapers were filled with articles about the great discovery. But Pluto, which was shrinking even before it was discovered, was shockingly little when Tombaugh first laid eyes on it, and it continued shrinking for half a century. Within one month, a report in the *New York Times* explained that the astronomy community was no longer certain about what Tombaugh had discovered:

> Doubts that the new member of the solar system discovered beyond Neptune is the ninth major planet predicted by Percival Lowell have been raised by computations of its orbit made at the Lowell Observatory. . . . It is now thought it may prove to be a unique asteroid or an extraordinary comet-like object.[1]

Such comments now look very prescient. By the time Christy discovered Pluto's moon Charon, in 1978, Pluto had shriveled to barely one five-hundredth the mass of the Earth. Pluto was not a massive planet like Uranus and Neptune, it was a disappointment, incapable of producing the nonexistent perturbative effects on the orbits of Uranus and Neptune that had led, in the first place, to the predictions by Lowell, Pickering, and others that Pluto existed.

Finally, in the 1990s, enticed by the discovery of the centaur Chiron in 1977, Dave Jewitt and Jane Luu opened up our eyes to a brand-new—to us—part of the solar system, the Kuiper Belt. Their work along with that of many others has demonstrated that Pluto shares a great deal in common with Ceres, in that Pluto is the biggest (currently known) object in the Kuiper Belt and is the biggest known of the remaining Plutinos. Its status as the biggest known object in the Kuiper Belt probably explains why Tombaugh found Pluto long before any other such objects were found and why Tombaugh himself failed to find any other trans-Neptunian objects.

So what is Pluto?

Dave Jewitt voiced his opinion on his Web page:

So, bluntly put, one has two choices. One can either regard Pluto as the smallest, most peculiar planet moving on the most eccentric and most inclined orbit of any of the planets or one can accept that Pluto is the largest known, but otherwise completely typical, Kuiper Belt object. The choice you make is up to you, but from the point of view of trying to understand the origin and significance of Pluto it clearly makes sense to take the second option. Pluto's eccentricity and inclination were pumped up along with the eccentricities and inclinations of the 25,000 other Plutinos (diameters > 100 km), probably driven by the radial migration of Neptune. The processes that shaped the orbits of the KBOs are the same ones that gave Pluto its prominent dynamical characteristics.

Some people see this as a demotion of Pluto from planet-hood. I think that it can reasonably be portrayed as a promotion. Our perception of Pluto has been transformed from a singularly freakish and unexplained anomaly of the outer solar system to the leader of a rich and interesting family of trans-Neptunian bodies whose study will tell us a great deal about the origin of the solar system.

So, we have discovered minus one planet and plus one Kuiper Belt. It seems like a fair trade to me.[2]

Astronomer Dan Green suggests that were it not for the political interests of Lowell Observatory in furthering the fame of their founder, staff, and observatory, Pluto would never have been considered a planet. Green uses a quote from Abraham Lincoln to amplify his feelings:

Getting recognition for an inflated categorization for political benefit of an individual or group (as Lowell Observatory did for Pluto in 1930) is not uncommon. Recently, Vermont politicians got Congress to recognize Lake Champlain as one of the "Great Lakes" so that Vermont could benefit from legislation directed at the Great Lakes.

Ian Ridpath [1978, *Astronomy* 6(12), 6] put it well when he recited Abraham Lincoln's riddle, in which Lincoln asked somebody, "If you call a tail 'a leg,' how many legs does a dog have?" The person offered,

"Five?" to which Lincoln replied, "No, four—calling a tail 'a leg' does not make it a leg!" To which Ridpath added: "If you call Pluto 'a planet,' how many planets does the sun have?"[3]

When schoolchildren learn the names of the planets, they often learn a mnemonic to remember nine, not eight, planets: My Very Earthly Mother Just Served Us Nasty Pizza. Indeed, when college students open their textbooks, they learn about nine planets—four earthlike planets, four gas giant planets, and the ice dwarf Pluto. Perhaps the next generation of textbooks will treat Pluto as a giant comet or a large Plutino, rather than a planet. But why? What is the rationale for treating any object as a planet or as not-a-planet?

We can draw a reasonable parallel between Pluto and Ceres. If we do not consider Ceres a planet but call it the largest asteroid, why do we call Pluto a planet rather than the largest Kuiper Belt object? Can Pluto be both? If it can, should Ceres be both also? Does the solar system really have ten planets, counting Ceres?

Making such a decision should not be an arbitrary activity. *Is Pluto a planet?* is a scientific question, not a matter of public opinion or a decision to be made by NASA or a panel of distinguished astronomers. Science moves forward at a pace dictated by progress in understanding, not by fiat or a majority vote of a committee.

The debate over the status of Pluto leads us naturally back to the question at the heart of this book: *What is a planet?* If we can answer this last question, if we can define a planet, perhaps we can decide whether Pluto (and Ceres and Quaoar) merit the designation *planet.* Let's now take a different approach, one that has little to do with the objects in our solar system. Let's investigate the question *What is a planet?* more generally. What qualities must an object possess in order to be called a planet?

14
Goldilocks

O

Not so long ago, at the end of the twentieth century, most astronomers would have agreed with three fundamental criteria that must be satisfied for an object to be considered a planet:

1. the object must be too small to generate, or to have ever generated, energy through nuclear fusion;
2. the object must be big enough to be spherical;
3. the object must have a primary orbit around a star.

We could summarize these three criteria by saying that a planet cannot be too big or too small, and it has to be in just the right place.

The first two of these criteria define a planet by what it is not. A planet is neither a star nor a brown dwarf nor a paper clip. Stars are enormous balls of hydrogen and helium gas that generate energy via the process of nuclear fusion. For most stars, for most of their lifetimes, the primary nuclear fusion reaction fuses four hydrogen atoms (actually, four protons) into one helium nucleus in a process called the proton-proton chain, thereby changing small amounts of mass into energy and powering the star.[1] Brown dwarfs are more massive than planets but less massive than stars. Brown dwarfs are unable to sustain nuclear fusion for more than a few tens of millions of years because they are not massive enough to start the proton-proton chain; however, at the lower temperatures and pressures they can generate in their cores, brown dwarfs can fuse the small amounts of deuterium (heavy hydrogen, containing one proton and one neu-

185

tron), lithium, beryllium, and boron that newly formed objects made of the dominant gases in the universe possess when they are born. Thus, brown dwarfs perform fusion for less than one hundred million years and then begin a process of cooling off that takes tens of billions of years.

Jupiter and Saturn could be classified as enormous balls of hydrogen and helium gas; however, Jupiter and Saturn are not massive enough to generate the enormous pressures and temperatures at their cores that are necessary in order to trigger any kind of nuclear fusion reactions; thus Jupiter and Saturn are neither stars nor brown dwarfs. Objects that are about seventy-five to eighty times more massive than Jupiter will become stars, and objects that are approximately thirteen times more massive than Jupiter (or four thousand times more massive than Earth) are expected to become brown dwarfs. We now know that an object can be too big to be a planet, and we have a good reason, based on the fundamental physics of the way the universe works, for drawing a dividing line between planets and larger objects.

Whereas the first criterion defines the upper limit for the mass of a planet, the second criterion defines a lower limit. So how big must an object be for us to consider it a planet? If we could place a paper clip or book or boulder or even a fifty-story apartment building in orbit around the Sun, we certainly would not call any of these objects planets, but why not? Again, physics. Planetary scientists have argued that for an object to be a planet, it must be big enough that its shape is determined by the force of gravity, not by the molecular and intramolecular forces that determine the shapes of smaller objects. Once an object is sufficiently massive that gravitational pressure overwhelms the molecular bonds of the materials that gave the object a particular shape, the object will take on the only shape that is possible for a slowly rotating object primarily under the influence of gravity (and perhaps subject to tidal effects that can also distort the object from perfect sphericity): a sphere. Having answered our question with a criterion—big enough that the shape of the object is controlled by gravity, i.e., spherical—we now must figure out how big that is: 1 or 10 or 100 or 1,000 km in diameter?

Figure 14.1: *Galileo* spacecraft image of asteroid 951 Gaspra (20 km along long axis). (Image courtesy of NASA/JPL-Caltech)

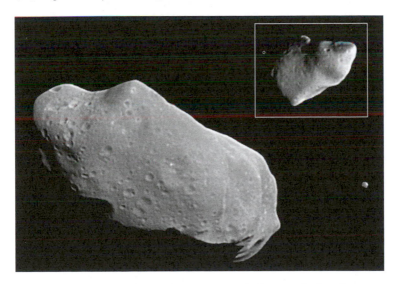

Figure 14.2: *Galileo* spacecraft image of asteroids Ida (36 × 14 km) and its moon Dactyl (small object to the right; inset). (Image courtesy of NASA/JPL-Caltech)

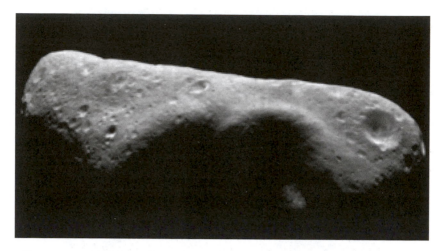

Figure 14.3: *NEAR* spacecraft image of asteroid 433 Eros (40 km × 14 km × 14 km). (Image courtesy of NASA/JPL-Caltech)

In fact, there is no simple answer. We cannot simply assign a size, since the self-gravity of an object is determined by a combination of the mass and the size; thus, a 100 km–sized object made of iron may take on a more spherical shape than a 100 km–sized chunk of water ice, since iron is five times denser (more mass in the same volume) than water ice. But we can take an empirical approach in order to determine a reasonable size above which objects take on a spherical shape. Let's look at objects in the solar system for examples of spherical and not-spherical objects.

NASA spacecraft have now obtained close-up images of the asteroids Gaspra, Ida, Dactyl, Eros, AnneFrank, and Mathilde as well as Mars's two moons, Phobos and Deimos, which are almost certainly captured asteroids. The images reveal that none of these objects are spherical. The biggest of these objects is only about 60 km across its longest axis. Planets, if they are to meet our criterion of sphericity, therefore must be bigger than this 60 km limit.

Astronomers also have obtained images of several large and small asteroids from ground-based telescopes. One of these, 216 Kleopatra, has a long axis that extends 217 km and a minor axis that is 94 km in length. Another, 511 Davida, extends about 320 km in its longest direction. Such objects obviously are not spherical.

Figure 14.4: *NEAR* spacecraft image of asteroid 253 Mathilde (61 km). (Image courtesy of NASA/JPL-Caltech)

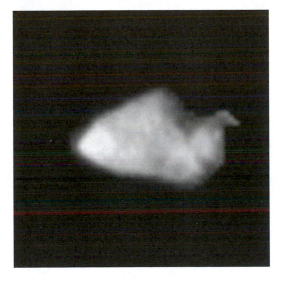

Figure 14.5: STARDUST spacecraft image of asteroid 5535 AnneFrank (6 km). (Image courtesy of NASA/JPL-Caltech)

Many moons are also odd-shaped, not spherical but not quite elliptical (or egg-shaped) either. Planetary scientists characterize these objects as triaxial, shaped like large, lumpy baking potatoes. Such objects are massive enough that gravity is beginning to work to assert some control over the shape; yet, the self-gravity of a triax-ial object is not quite strong enough to turn a large, lumpy potato

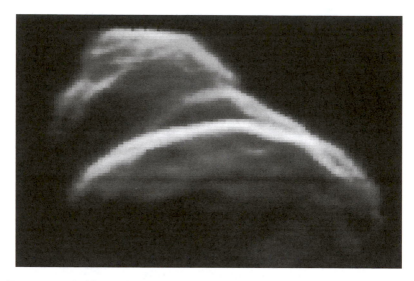

Figure 14.6: Goldstone (California) radar image of asteroid 4179 Toutatis (1.92 km × 2.29 km × 4.60 km), obtained when Toutatis was only 0.024 AU from Earth. (Image courtesy of NASA/JPL-Caltech)

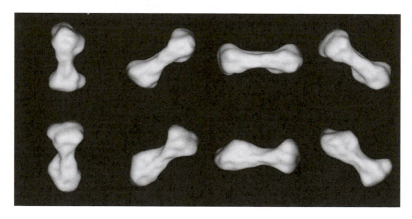

Figure 14.7: Arecibo (Puerto Rico) Observatory radar image of asteroid 216 Kleopatra (217 km in length; 94 km in width), obtained when Kleopatra was 171 million km from Earth. (Image courtesy of NASA/JPL-Caltech)

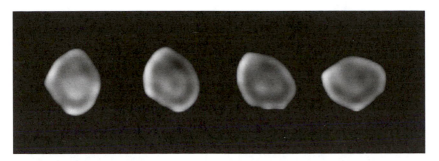

Figure 14.8: W. M. Keck (Hawaii) Observatory images of asteroid 511 Davida (320 km along its longest axis), revealing the asteroid as it spins counterclockwise about one-quarter turn over a one-hour time period (the time series begins with the leftmost image), as seen from above its north pole. These images clearly reveal the oblong shape of Davida, which astronomers had suspected for a long time, based on brightness variations of this asteroid. (Image courtesy of the W. M. Keck Observatory, William Merline of the Southwest Research Institute, and Al Conrad and Christopher Dumas of the Jet Propulsion Laboratory)

into a sphere. Saturn's moons Pandora (114 km × 84 km × 62 km), Epimetheus (144 km × 108 km × 98 km), and Janus (196 km × 192 km × 150 km) all are good examples of triaxial objects. Even Hyperion (410 km × 260 km × 220 km), which is twice as large along most of its axes as the smaller moons, clearly is not spherical. Neptune's moon Nereid, whose size is not well measured but which has a long axis diameter of 340 km, like Hyperion, is most certainly not spherical. We know, then, that objects with sizes of a few hundred kilometers become triaxial but may not become spherical.

How big must an object be to be spherical? The Earth's moon is spherical. Jupiter's four moons discovered by Galileo are all large spherical objects, the smallest one being Europa with a diameter of 3,138 km; but these moons are quite big compared to the small moons of Saturn and Neptune and the few asteroids that have diameters of a few hundred kilometers. What intermediate-sized objects do we have in the solar system? Triton is a bit smaller, with a diameter of 2,700 km, and is clearly spherical. As illustrated in the figures, working our way downward in size from the Moon to Pluto and then to the moons of Saturn and Uranus, we find all of these, all the way down to Enceladus (diameter = 498 km), are spherical.

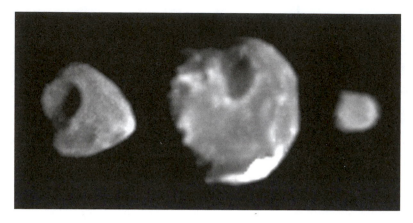

Figure 14.9: Three of Jupiter's small moons (*from left to right*): Thebe (55 × 45 km), Amalthea (270 × 165 × 150 km), and Metis (60 km), in images obtained by NASA's *Galileo* spacecraft. (Image courtesy of NASA/JPL-Caltech)

We do not have detailed images and therefore have not actually measured the three-dimensional shapes of the asteroids Ceres (diameter = 925 km), Pallas (diameter = 583 km), and Vesta (diameter = 555 km), and the ten largest known Kuiper Belt objects, Varuna (diameter = 760–1,040 km), Ixion (diameter = 900–1,230 km), 2002 AW_{197} (diameter = 700–1,010 km), Quaoar (diameter = 1,300 km), 2002 TC_{302} (diameter = 1,200 km), 2003 EL_{61} (diameter = 1,500 km), 2003 UB_{313} (diameter = 2,700 km), Sedna (diameter = 1,500 km), Orcus (diameter = 1,400 km), and 2005 FY_9 (diameter = 1,500 km). All of these objects, however, are substantially bigger than Enceladus, and many are much bigger than Tethys (diameter = 1,060 km). The smallest of them (Pallas, Vesta) are rock and iron-rich asteroids, not balls of ice, and so are much denser than an icy moon of comparable size, like Enceladus. Without a doubt, therefore, we can assume that all of these objects are spherical.

The biggest, well-observed solar system object below the size limit set by Enceladus is the Uranian moon Miranda. Miranda has a diameter of 472 km and is fairly well characterized as very nearly spherical; Miranda, though, has some rough edges. Some of Miranda's mountains, which are scars left by giant meteor impacts billions of years ago, reveal topographic relief that Miranda's gravity cannot quite pull back down. Saturn's Mimas is a little smaller, with a diam-

Figure 14.10: *Galileo* spacecraft image of Earth's moon (diameter = 3,476 km). (Image courtesy of NASA/JPL-Caltech)

Figure 14.11: *Galileo* spacecraft image of Jupiter's moon Europa (diameter = 3,138 km). (Image courtesy of NASA/JPL-Caltech)

Figure 14.12: *Voyager 2* image of Neptune's moon Triton (diameter = 2,700 km). (Image courtesy of NASA/JPL-Caltech)

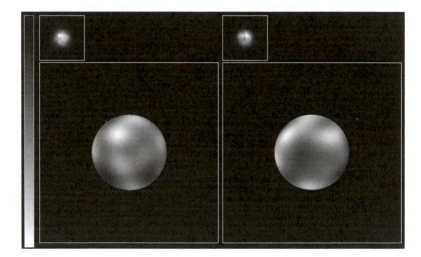

Figure 14.13: Images of Pluto made in blue light using the Faint Object Camera on the Hubble Space Telescope in 1994. The two smaller inset pictures at the top are actual images from Hubble. Each square pixel (picture element) is more than 160 km across. At this resolution, Hubble discerns roughly twelve major "regions" where the surface is either bright or dark. The larger images (*bottom*) are from a global map constructed through computer image processing performed on the Hubble data. The tile pattern is an artifact of the image enhancement technique. (Images courtesy of Alan Stern of the Southwest Research Institute, Marc Buie of Lowell Observatory, NASA, and ESA)

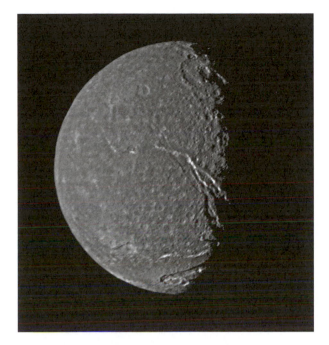

Figure 14.14: *Voyager 2* picture of Uranus's largest satellite Titania (diameter = 1,578 km). (Image courtesy of NASA/JPL-Caltech)

Figure 14.15: Cassini picture of Saturn's satellite Rhea (diameter = 1,528 km). (Image courtesy of NASA/JPL-Caltech)

Figure 14.16: *Voyager 2* picture of Uranus's moon Oberon (diameter = 1,523 km). (Image courtesy of NASA/JPL-Caltech)

Figure 14.17: *Cassini* picture of Saturn's moon Iapetus (diameter = 1,430 km). (Image courtesy of NASA/JPL-Caltech)

Figure 14.18: *Voyager 2* picture of Uranus's moon Umbriel (diameter = 1,170 km). (Image courtesy of NASA/JPL-Caltech)

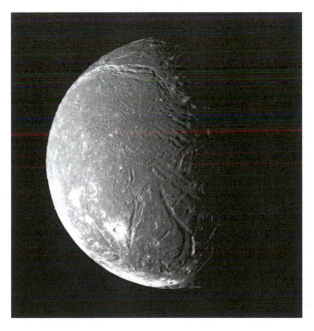

Figure 14.19: *Voyager 2* picture of Uranus's moon Ariel (diameter = 1,158 km). (Image courtesy of NASA/JPL-Caltech)

Figure 14.20: Computer-generated images from brightness changes of Pluto and Charon, using data from 1954 through 1986. Pluto is on the left, Charon (diameter = 1,300 km) on the right. This view shows the side of Pluto that faces away from Charon and the side of Charon that faces Pluto. (Image courtesy of Marc W. Buie/Lowell Observatory)

Figure 14.21: *Cassini* picture of Saturn's moon Dione (diameter = 1,120 km). (Image courtesy of NASA/JPL-Caltech)

Figure 14.22: *Voyager 2* picture of Saturn's moon Tethys (diameter = 1,060 km). (Image courtesy of NASA/JPL-Caltech)

Figure 14.23: *Cassini* picture of Saturn's moon Enceladus (diameter = 498 km). (Image courtesy of NASA/JPL-Caltech)

Figure 14.24: *Voyager 2* picture of Uranus's moon Miranda (diameter = 472 km). (Image courtesy of NASA/JPL-Caltech)

eter of 398 km. As with Miranda, we might describe Mimas as nearly spherical, albeit with one enormous crater and an impressive amount of topographic relief. We might conclude, fairly safely then, that 400 km is a good, though only approximate, boundary for sphericity. If an object is larger than 400 km, it will be best described as spherical or very close thereto; if the object is smaller than this limit, the object is likely better described as triaxial.

This 400 km limit is not perfect or even one-size-fits-all. For example, if one object is made of rock and iron and a second object of the same size is made of ice, the self-gravity of the rock and iron-rich object will be three to five times greater than that of the ice ball and the resulting pressures in the core of the object will be

Figure 14.25: *Cassini* picture of Saturn's moon Mimas (diameter = 398 km). (Image courtesy of NASA/JPL-Caltech)

comparably greater. Since an object's own gravity is the force that squeezes it into a spherical shape, the more massive, rock and iron-rich object is more likely to be spherical than is the less massive, icy object.

In the outer solar system, the moons—like Miranda and Mimas—are dominantly made of ice, and so our 400 km limit was found for relatively low-mass (or low-density) objects. We might reasonably conclude that a rocky, iron-rich asteroid would be spherical if it had the same mass as a 400 km ball of ice. Thus, a denser asteroid-like object might be as small as 275 km in diameter and still be spherical. Most asteroids are well below our 275 km limit; however, in addition to the big three, Hygeia (443 km) and Cybele (311 km) may be above the 275 km size threshold for asteroids. On the other side of the threshold, images of Juno (249 km) clearly show that it is potato-shaped.

Our third criterion is fairly simple: the primary orbit of an object must be around a star. The Moon, for example, is big enough to be spherical but too small to be a star; however, although the Moon orbits the Sun, it is carried around the Sun by the Earth as the Earth orbits. The primary orbit of the Moon is around the Earth. Thus, the Moon is a moon, not a planet. Jupiter's moon Ganymede is bigger than Mercury; however, since the primary orbit of Ganymede is around Jupiter while that of Mercury is around the Sun, Mercury is called a planet while Ganymede is relegated to the lesser status of moon. Similarly, Triton and Charon fail the "primary orbit around a star" criterion, as they are moons of Neptune and Pluto, respectively.

What about Pluto? Pluto meets all three of our constraints. Though it is smaller than many moons, it is not too big and not too small and it orbits the Sun. By our three-point definition, Pluto is most certainly a planet.

Of course, our three-point definition also would turn Ceres, Pallas, Vesta, Hygeia, and Cybele, among the asteroids, into planets. Furthermore, several of the known Kuiper Belt objects, including Varuna, Ixion, 2002 AW$_{197}$, Quaoar, 2002 TC$_{302}$, 2003 EL$_{61}$, 2003 UB$_{313}$, Sedna, Orcus, and 2005 FY$_9$, meet our conditions, and almost certainly more yet-to-be-discovered KBOs will also qualify as planets. At a minimum, then, the Sun has twenty-four planets: the eight well-known planets, from Mercury to Neptune, the controversial Pluto, and the fifteen objects just listed that are usually labeled asteroids or KBOs.

Now, no one has argued in recent times that the Sun has ten planets, let alone twenty-four; yet, we have not found a reasonable set of conditions that we can use to define *planet* that would include the traditionally accepted nine planets but would exclude Ceres, Quaoar, and their ilk. Perhaps, then, we can improve and provide a more complete definition of *planet* by imposing additional criteria.

TESTING OTHER POSSIBLE PLANET-DEFINING CRITERIA

We can imagine other constraints that might help us define a planet in a physically meaningful way.

Let's try one: *Can we assert that a planet must have an atmosphere?*

Venus, Earth, and Mars have solid, earthlike surfaces with gaseous atmospheres (as does Saturn's moon Titan). Jupiter, Saturn, Uranus, and Neptune lack solid surfaces and can be thought of as giant balls of gas, nearly 100 percent atmosphere. Pluto appears to meet this criterion for at least a small fraction of its orbit, so we might need to amend this constraint: a planet must have an atmosphere for at least part of its star-centered orbit. Now we can include Pluto. Mercury, however, never has an atmosphere. If we accept this requirement, we will successfully exclude the largest asteroids and Kuiper Belt objects, but we also will eliminate Mercury. Tradition makes this unacceptable. Furthermore, since a moon can have an atmosphere, such a consideration would be nonunique and lacks a strong basis in physics. The "must have an atmosphere" criterion clearly fails.

Let's try another: *Can we assert that a planet must have a moon?*

The Earth has a moon. Mars has two moons now, but did not always, since Mars's moons are almost certainly captured asteroids. The four giant planets and Pluto have moons. Mercury and Venus, however, do not have moons. This "must have a moon" requirement would drop Mercury and Venus from the list of planets and make Mars questionable. Did Mars become a planet only after it captured its moons from the asteroid belt a few hundred million years ago? If a Mars-sized object had not crashed into Earth 4.5 billion years ago, and if the Moon therefore had never formed from the collisional detritus, would the Earth not be a planet? No, this "must have a moon" criterion will not help us better define a planet.

Can we demand that a planet must have liquid water? No; in our solar system only the Earth and perhaps Mars satisfy this requirement.

Can we require that a planet must have life or be capable of supporting life? No; such a requirement would reduce the number of planets around the Sun to only one, as far as we know, or perhaps two if bacteria-like life exists on Mars. And did the Earth become a planet only after life formed?

Can we impose a constraint that planets receive more energy (from outside) than they generate (from internal sources)? No; the Earth meets this criterion, but Jupiter, Saturn, and Neptune fail. These three planets are giant balls of gas that are cooling off over the eons. As they cool off, they contract; as they contract, gravity converts a form of energy called potential energy into heat, which is then radiated into space, and these three planets all generate more energy than they receive from the Sun.

Can we insist that a planet must have a solid surface? No; none of the gas giants—Jupiter, Saturn, Uranus, and Neptune—have solid surfaces.

Can our definition require the presence of a magnetic field? No; such a criterion would eliminate Venus and perhaps Pluto. In addition, Mercury and Mars appear to have only remnant magnetic fields (i.e., they do not continually generate their magnetic fields), so eventually, or perhaps already, Mercury and Mars would change from planets into nonplanets.

Can we demand that little, spherical objects are too little to be planets? A diameter of 3,000 km would isolate Mercury above the boundary and Pluto below the boundary. But why not 2,000 km or 4,000 km? This criterion is completely subjective and provides nothing meaningful in terms of physics that would distinguish "big enough" planets from "too little" nonplanets.

Apparently, all the obvious requirements beyond the three definitional constraints we started with have fatal flaws. The original three rules are the only ones that make any sense whatsoever for defining a planet, yet even those three do not uniquely distinguish the traditional nine planets from nearly a dozen other solar system objects.

CAN WE DEFINE A PLANET BY WHAT IT IS NOT, RATHER THAN BY WHAT IT IS?

The three widely accepted rules for defining a planet that we started with are such that the well-known planets from Mercury to Neptune are all included. But as we have also recognized, these rules also demand that Pluto and at least a small handful of Plutinos and Kuiper Belt objects as well as Ceres and a few other asteroids be called planets. If we want another criterion that distinguishes the eight largest spherical objects orbiting the Sun from the dozen or more smaller but nevertheless spherical objects also in helio-centric orbits, *we could insist that planets are not part of rings or belts around a star.* Thus, Ceres and the largest asteroids could be con-sidered merely the largest known objects in one ring around the Sun, while Pluto is the largest object in the Plutino ring, and 2003 EL_{61} is the largest object currently known in the main part of the Kuiper Belt.

This criterion does appear to clearly distinguish the eight largest spherical bodies in our solar system from the smaller spherical bod-ies. On the other hand, at a distance of 5.2 AU from the Sun, located at positions always centered sixty degrees ahead of and sixty degrees behind Jupiter, are clouds of objects called Trojan asteroids. These asteroids cluster around unique locations known as the L4 and L5 Lagrange points. At these two locations, the competing gravita-tional pulls from the Sun and Jupiter act to lock these objects into stable orbits with nearly the same orbital period as Jupiter. By the end of 2005, astronomers had identified 1,844 objects clustered at the Jovian L4 and L5 positions. These objects span a range in longi-tude that covers about a forty-five degree arc centered on each of the L4 and L5 points; that is, the Jovian Trojan asteroids fill about 25 percent of Jupiter's orbit. Jupiter, therefore, is the largest object in the Jovian ring. To complicate matters further, with the discovery of Neptune's first four Trojan asteroids, objects 2001 QR_{322}, 2004 UP_{10}, 2005 TN_{53}, and 2005 TO_{74}, Neptune joined Jupiter and perhaps Mars (seven unconfirmed Trojans, the first discovered in 1990 and the rest since 1998, with two Trojans at L4 and five at L5) as planets with Trojans.[2] If the largest object in a ring or belt is not a planet, we must remove from our list of planets Ceres and Pluto and also Jupiter, Neptune, and Mars. Such a definitional criterion is unacceptable. Rather than use these observational results to narrow the boundaries for planets, we should use the "belt-embedded" constraint to enhance our understanding of the conditions in which certain planets are able to exist.

We could add a caveat to the belt-embedded criterion: we could distinguish wide, filled rings like the asteroid and Kuiper belts from narrow, unfilled rings that have one dominant planet in addition to smaller objects only at the dominant planet's L4 and L5 points. But this now overcomplicated criterion that says that objects that are the largest bodies in unfilled rings are planets but the largest bodies in filled rings are not planets is a flawed means for determining whether an object is a planet, as it does not make reference to the physics of the object itself.

ORPHANED PLANETS

Even the three widely accepted criteria for defining planets are not unassailable. For example, imagine that a star with ten planets formed as part of a cluster of dozens or even thousands of other stars, in a star-forming environment like the Orion Nebula. In some of these clusters, the average distance between newborn stars is only a few thousand AU.

Within a young star cluster, the stars orbit a common center, sometimes passing very close to other newborn stars. When one star passes close to another, the gravitational attraction of the approaching star increases as the distance between the stars decreases. Any planets orbiting either star will feel, for an extended period of time while the two stars are close together, a gravitational force tugging them out of their orbits around their home star. These tugs will change the orbits of the planets in the same way that a spacecraft

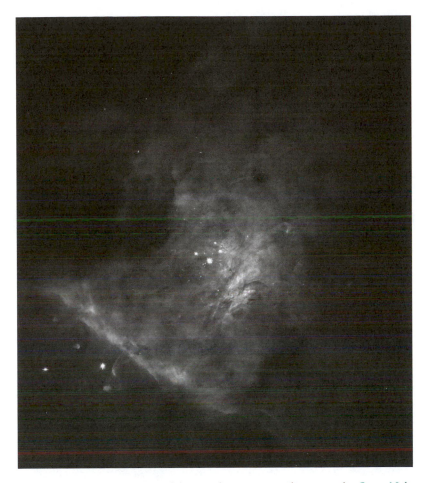

Figure 14.26: A mosaic image of the star-forming region known as the Orion Nebula, obtained with the Hubble Space Telescope. This region of the Orion Nebula spans about 2.5 light-years and is located about 1,500 light-years from Earth. Many newborn stars are visible in this image. (Image courtesy of C. Robert O'Dell, Vanderbilt University, and NASA)

sent by NASA engineers into an orbit around the Sun but which passes close to Venus will be slingshot by Venus into a new orbit, perhaps toward Jupiter or Saturn. Over time, such close encounters in a dense cluster of newborn stars or even among planets and stars in a multiple star system will disperse the cluster, sending individual stars out into the galaxy, far from their birthplaces. These close encounters also may rip planets right out of their orbits and toss them into interstellar space.

The planets around our star include a few in small orbits bound tightly to the Sun by virtue of the small separation between the Sun and the inner planets. Some of the outer planets, however, have orbital semi-major axes of a few tens of AU and are much more loosely bound by the Sun's gravity. Now imagine a young, Sun-like star in a dense cluster of newborn stars in which the average separation between stars is only 1,000 AU and stars randomly pass within 50 to 100 AU of each other in the first 10 million years after they are born. During one close encounter with a nearby star, the young Sun-like star and its neighbor come within only 50 AU of each other. At the moment when these two stars are separated by only 50 AU, we imagine that an outer planet (of either star) happens to be at the point in its orbit that places it exactly 25 AU from its parent star and exactly in between the two stars. At this moment, this unfortunate planet feels the gravitational attraction of both stars at equal strength. The effect of being pulled in two different directions by the two stars is such that the planet no longer knows which star to orbit. In fact, this planet follows a new orbital path controlled by both stars, and as the stars pass each other and separate, the planet follows its own path out into interstellar space, no longer in orbit around either star.

Such close encounters regularly tear members of binary star systems loose to become single stars inside these dense star clusters. So, almost certainly, such planet-freeing encounters can occur even more readily than close encounters that disrupt binary star systems. Thus, after such a close encounter, the formerly outer planet finds itself ripped free from its orbit, orphaned in interstellar space. Would this formerly outer planet no longer be a planet because it no longer orbited a star?

While the likelihood of another star tearing planets loose from the Sun is extremely remote at this time, one can easily imagine that in young star clusters, gravitational close encounters occur often and with regularity. Only planets in the most tightly bound orbits are likely to survive until the young star cluster disperses after perhaps a hundred million years. Consequently, if we could search within clusters of newborn stars, we would very likely, find hundreds or perhaps thousands of objects that we would classify as planets were they orbiting stars; yet these objects would be orphans, torn loose from their moors around their parent stars. We might coin a variety of names for these objects—ripped-free planets, liberated planets, orphaned planets—but it is clear that the discoveries of modern astrophysics will further complicate our ability to easily define an object as a planet.

FREE-FLOATING PLANETS

Planets torn free from where they were born may not be our only problem with our original three criteria. We have assumed that planets are supposed to be in orbit around stars, though we have accepted the possibility that some planets might be liberated from their orbital bonds by the gravitational effects of nearby stars. But what if planets are not always supposed to be in orbit around stars? Might a planet form all by itself?

Since some brown dwarfs are failed stars, might some failed stars be so small that they are also failed brown dwarfs? Can the star-making process fail so early that the object created is below the brown dwarf limit? If a future explorer or astronomer discovered an object like Jupiter that was unattached to a parent star, and if this explorer somehow were able to prove that this object had formed on its own, never having been ripped loose from a parent star, would this object be a planet? It would not be big enough for nuclear fusion of any kind. Most certainly, though, it would be spherical. However, it would not have a primary orbit around a star and it would not have been ripped out of a stellar orbit, as a liberated planet. Might such free-floating planets, ones that simply formed via a different process, exist? If such free-floaters could form in this manner, then the entire constraint "the primary orbit is around a star" might be completely unreasonable from the start.

Astronomers' models for making planets assume that planets form in disks of gas and dust around newborn stars. In these models, the process of making a planet is fundamentally different from the process by which stars form. So, in order to better assess whether a planet could form by itself, we need to better understand the star formation process.

Stars form by a process known as gravitational collapse. Interstellar clouds of gas and dust are normally too warm to permit stars to form. "Too warm," in this case, means that the molecules in the interstellar cloud have enough energy and are moving fast enough to resist being pulled closer together by gravity. Every now and then, though, a small piece of a giant cloud will find a way to cool off. As the cloud fragment cools, the self-gravity of the cloud overwhelms the thermal energy of the cloud; when this occurs, the cloud contracts. Gravity continues to pull all the material in the cloud toward the center, eventually making a very dense and massive object. If this object is massive enough, equivalent to more than seventy-five or eighty Jupiters put together, the temperature at the center of this collapsing object will reach a few million degrees. In the core, nuclear fusion reactions will occur in which hydrogen nuclei (protons) combine to form helium nuclei. Each nuclear fusion reaction converts a small amount of mass into energy. The release of energy from nuclear fusion reactions in the core heats up the inside of the star and halts the gravitational contraction. Such an object, a star, then continues to fuse light nuclei into heavier nuclei for millions or even billions of years.

If the object is a bit less massive, with between thirteen and seventy-five Jupiters of material, some nuclear fusion reactions, especially of the heavy isotope of hydrogen known as deuterium, may occur. But they either will not continue for long or will never liberate enough energy to halt the slow gravitational contraction of this object. Such objects are known as brown dwarfs.

From decades of observing stars, astronomers were fairly certain, well into the 1980s, that the star formation process was good at making stars but not so good at making smaller objects. The smallest, lowest mass stars had masses of about 8 percent that of the Sun, or the equivalent of 80 Jupiter masses. Apparently, the process that was good at making stars was not as good at making substellar objects, the hypothesized brown dwarfs. Discoveries made over the last twenty years, however, have destroyed that paradigm. Brown dwarfs exist in abundance. Some brown dwarfs orbit stars, acting as members of binary star systems in which one of the two components was not quite big enough to become a star. Others may be free-floaters.

The first discoveries of brown dwarfs and extrasolar planets suggested the possibility of a brown dwarf "desert," encompassing the mass range from about 10 to 50 Jupiter masses. That is, astronomers were finding brown dwarfs at the very high mass end of the brown dwarf range (masses just less than seventy-five Jupiters) but no smaller ones. For a few years, it appeared that the process that makes stars failed at masses just below the minimum mass for stars, whereas the process that makes planets was unable to make planets much bigger than about ten Jupiters. The brown dwarf "desert," though, appears to have been a mirage, with brown dwarfs having been found covering the entire range of masses from super-Jupiters to almost-stars. How small, we must ask, is the smallest possible object that can be made through the "star-forming" mechanism? Could the star-forming process truncate early, leaving behind an object with the mass of Jupiter? If so, is this object a planet?

Since we most likely can never know the process by which a free-floating object formed, we would have a very difficult time applying such a criterion to evaluate whether a 10-Jupiter-mass object should be considered a failed star or a large planet. Almost certainly, we should not include formation criteria in our definition of a planet.

Beginning in 1998, a number of claims have been put forward asserting that *free-floating planets* had been found. Thus far, none of them have withstood scientific scrutiny. Nevertheless, it is clear that free-floating objects with masses comparable to those expected for giant Jupiters could exist, even if we have not yet found and identified one. Assuming that someday one is found, how will we categorize such an object? Would it be a planet?

PULSAR PLANETS

Yet another complication for our understanding of planets arose with the discovery in 1991 by Alex Wolcszan of an object orbiting a pulsar whose mass is comparable to that of the Earth. By 1993, three "planets" with masses of 4.3, 3.9, and 0.02 *Earth* masses, located at distances of only 0.19, 0.36, and 0.46 AU from the central object, had been discovered around the pulsar PSR 1257+12. More recently, a claim has been made for a fourth planet, with a mass of 100 Earth masses and at a distance of 40 AU from the central pulsar, as part of the PSR 1257+12 system.

Immediately upon the report of the discovery of the planets around PSR 1257+12, planetary scientists asked, *Can we have planets orbiting pulsars?* Pulsars are dead stars, the remnants of stars much more massive than the Sun. Such stars die violent, energetic deaths: their precursor stars end their lives as regular stars by exploding as supernovae. In the cataclysmic explosion that produced the pulsar—so named because the stellar remnant beams its remaining light output out of its rapidly spinning north and south magnetic poles and so looks to us to be giving out light pulses, although it is not—most of the mass of the star is driven away from the star at speeds of tens of thousands of kilometers per second. The tremendous outflow of mass and energy from the exploding star, combined with the shock waves generated by the explosion, might blast apart and even vaporize any planets that may have been orbiting the star prior to the explosion. Thus, according to this scenario, the objects now in orbit around the pulsar, objects now referred to as "pulsar planets," must have formed after the star died. Alternatively, in the aftermath of the supernova explosion, the soon-to-be pulsar may have been driven into and through its own stellar companion. During this event, the pulsar might collect enough mass from the companion to form a new planetary system.

How pulsar planets form, no one yet knows. While these objects may be very strange, perhaps composed entirely of carbon produced in abundance in the exploding star, and while such planets are among the least likely sites in the universe for harboring life because of the intense, deadly radiation produced by the nearby pulsar, the question for us is simply, *Are "pulsar planets" planets?* They are big enough but not too big. The central object was a star but is no longer. If these planetary mass–sized objects formed at the moment of the death of the star, rather than at the time of birth of the star, are they disqualified as planets? In other words, does the time of formation of the planet-sized object matter?

GLOBULAR CLUSTER PLANETS

In July 2003, Steinn Sigurdsson added another strange planet to our collection. In 1992, astronomers first began to suspect that the pulsar PSR B1620–26 had a 2.5-Jupiter-mass planet located 23 AU from the central object—about the distance of Uranus from the Sun—in a 100-year orbit. PSR B1620–26 is located in a dense cluster of more than one hundred thousand ancient stars, known as Messier 4, at a distance of more than 7,200 light-years from Earth. In the 1980s, researchers realized that PSR B1620–26 had a white dwarf companion. Sigurdsson's work indicates that the planet was born orbiting the white dwarf, nearly 13 billion years ago, long before the star that became the white dwarf died. At some time long after the star and planet formed, the now–white dwarf passed so close to the neutron star that it, along with its planet, was captured into orbit around PSR B1620–26. Now, this super-Jupiter orbits what is now a central binary star, made up of the pulsar and the white dwarf.

When this super-Jupiter formed 13 billion years ago, the then-young stars in the Milky Way galaxy had not yet had a chance to create many atoms of carbon, oxygen, nitrogen, iron, or any other elements heavier than hydrogen, helium, and lithium (these lightest elements formed in the Big Bang itself, whereas the heavier elements are forged in the nuclear furnaces inside stars). Without any heavy elements available to form grains of dust and then planetesimals, this giant planet must have formed in the same way stars and brown dwarfs form, through gravitational collapse rather than through the collisional process that most planetary scientists believe is the formation mechanism for most planets and for the first stage of the formation of giant planets like Jupiter. Thus, this globular cluster planet raises yet another question about planets: *Does the mechanism by which a planet forms matter?*

ELLIPTICAL ORBIT PLANETS

Some planetary scientists have taken the position that planets form in disks around stars; such a formation location is consistent with our definition that planets have primary orbits around stars. If planets form in disks, such planets most likely will have orbits that are very nearly circles, unless they are very low-mass objects like Pluto that can get pushed around by the big planets in the disk. If we find a giant planet in a very elliptical orbit, then it may have formed in the same way that stars and brown dwarfs form, through the collapse of a large parcel of gas by the action of gravity. Do planets in highly elliptical orbits exist?

Since 1995, astronomers have identified more than one hundred objects in orbit around normal stars whose masses are below the brown dwarf limit. Presumably, all these objects are planets.[3] In fact, almost everyone would agree that almost all of these objects are planets. However, a few of them have orbits that are far more elliptical than even Pluto's orbit. For example, the planet known as HD 177830 b has an eccentricity of 0.43 and an orbital semi-major axis identical to that of the Earth, 1.0 AU. But whereas the Earth stays very nearly at exactly 1.0 AU at all times, HD 177830 b sometimes gets as close as 0.57 AU (closer than Venus is to the Sun) and other times as far as 1.43 AU (farther than Mars gets at closest approach to the Sun) from the star it orbits, HD 177830. *Is such an orbit reasonable for a planet?* Can a planet with such an elliptical orbit have formed in a disk rather than through the gravitational collapse mechanism by which stars are made? And does it matter? If we are to set a minimum allowable eccentricity for the orbit of a planet,

214

what would that be, who would decide, and what would the basis be for the critical cutoff value?

Some claims have been made that planets in highly elliptical orbits could only have formed as failed binary stars. Others have suggested that the gravitational interaction between two giant planets or between one giant planet and the disk in which it forms could generate a highly elliptical orbit.

The record for elliptical orbit planets so far is held by the companion to the star HD 80606, with an eccentricity of 0.927; however, nearly half of the known extrasolar planets have orbits far more elliptical than Pluto's orbit. Perhaps the orbits of most of the planets in our solar system are unusually circular. Perhaps the shapes of orbits tell us nothing about the circumstances through which a planet formed. If nothing else, the high eccentricities of the orbits of many extrasolar planets suggest that there is a great deal we do not yet understand about how planets form.

COULD A MOON BE A PLANET?

We know of many objects that would be called planets if their primary orbits were around stars. Some of these objects may once have had such orbits long ago; others might have formed on their own or formed around objects that are technically not stars, since they are not carrying out nuclear fusion reactions. We might even accept our labels of orphaned planets and free-floating planets and pulsar planets as additions to our lexicon. Should we then discard the constraint that planets must orbit stars or stellar remnants like neutron stars? And if we do, then what do we call Ganymede, Callisto, Europa, Io, Titan, Triton, Charon, and Earth's Moon?

The planet hunters Geoff Marcy and Paul Butler have defined a planet as follows: "A 'planet' is an object that has a mass between that of Pluto and the deuterium-burning threshold and that forms in orbit around an object that can generate energy by nuclear reactions."[4] The Marcy and Butler definition, which ignores formation processes but specifies formation location, excludes free-floaters and orphaned and pulsar planets, and sets the size of Pluto as an arbitrary minimum mass for a planet. But by this definition, Triton, which almost certainly *formed* in orbit around the Sun and was subsequently captured into orbit around Neptune, is a planet. Is this satisfactory?

What if, in another planetary system, we discover an object the size of the Earth orbiting a gas giant planet like Jupiter at a distance of 1 AU from the parent star. What if this Earth-sized object has oceans of liquid water, an atmosphere rich in oxygen, and an abundance of aerobic, multicellular life forms. Would we call this life-bearing, Earth-sized object a moon? A moon-planet? Or would we call this newly discovered object a member of a binary planet system?

REFOCUSING

Even if we keep our third criterion, which we have seen may not be such a good idea, we still are left with more than twenty planets in our own solar system unless we impose, like Marcy and Butler, an unphysically meaningful and arbitrary cutoff with Pluto. We could reinvigorate the designation *minor* planets,[5] and we could accept the historical constraint that Ceres and all objects of similar size or smaller are minor planets. Of course, in the same way that we are accepting liberated planets and free-floating planets as planets, we would be accepting minor planets as planets. This does not work as well, since most of the objects tracked by the Minor Planet Center fail to meet our minimum size requirement. As the terms are used by astronomers, most minor planets are too small to be *planets*, and therefore the term *minor planet* has come to mean something altogether different from what we mean by the term *planet*. As it is used, *minor planet* means an object in orbit around the Sun that is too small to be considered a planet.

For lack of a better term, we could coin another term for Ceres and those other minor planets that meet the physical criteria for planets but fail the historical criterion (c. 2005) for being planets. This would reduce the number of major planets to nine again.

Pluto, by this reckoning, would remain a planet. But this is a very unsatisfactory solution, with no basis in astronomy, physics, or even sociology.

We know that Pluto earned its status as a planet by accident. The predictions that led to Pluto's discovery were completely wrong. As careful as his mathematics may have been, Percival Lowell's work was heavily biased by his desire to find a ninth planet, so he attributed accuracies to observations of Uranus and Neptune far beyond those merited by the data. Lowell *created* the idea of Pluto since, as we now know, the data could not and did not reveal Pluto. Had Lowell recognized and understood the limited accuracy of the observational data at his disposal, he never would have begun his observational search for Planet X, the staff at Lowell Observatory would not have hired Clyde Tombaugh, and Pluto might not have been found for another half century. By the time Pluto would have been found, the discoveries of Chiron, Pholus, and the entire Kuiper Belt would have followed in rapid succession. Alan Stern's predictions that thousands of Plutos must once have existed in order to explain the existence of Triton and Charon also would have followed closely on the heels of the discoveries of a few odd objects in the outer solar system. Most likely, Pluto would have been treated in the same way as Ceres: Pluto would be called the largest (currently known) object in the Kuiper Belt (or perhaps the Plutino Belt), just as Ceres is the largest object in the asteroid belt. Ceres and Pluto both meet our criteria for being planets, yet both are parts of large swarms of objects orbiting the Sun in their respective parts of the solar system. Each is a dominant object in one of the two or perhaps three major rings presently known around the Sun. Had the century of discoveries of KBOs unfolded this way, we probably would not now call Pluto a planet. History, of course, did not play out this way. By happenstance, thanks to persistence and the then-unknown existence of the Kuiper Belt, Clyde Tombaugh was hired to look for a nonexistent Planet X and, instead, found Pluto. So Pluto is labeled a planet.

However, the historical pattern of discovery is not astrophysics, and no matter how or why Pluto was discovered, our definition of *planet* should be based on physical, not historical, criteria. As we have seen, only three criteria exist that are based in physics and that make consistent sense: planets are objects that orbit stars and that are not too big (never massive enough to generate energy through nuclear fusion) and not too small (shape is determined by gravity, so object is spherical). And if we use physics, not arbitrariness, to determine the limiting size at which gravity controls the shape of an object in orbit around a star, Pluto is a planet. That is our answer: *Pluto is a planet.*

And so is Ceres. Rewriting history will not change the fact that no criteria based on physical or astrophysical principles are sufficient to distinguish Pluto and Ceres from Mercury and Jupiter. So our answer must also include: *and so are Ceres, Pallas, Vesta, Hygeia, Cybele, Varuna, Ixion, 2002 AW$_{197}$, Quaoar, 2002 TC$_{302}$, 2003 EL$_{61}$, 2003 UB$_{313}$, Sedna, Orcus, and 2005 FY$_9$.*

In trying to understand whether Pluto merits being called a planet, in trying to understand what a planet is, we have to keep in mind that science is an enterprise done by people who work in the context of their times. Some great discoveries are made by accident, others by purposeful effort, and a few by dumb luck. As our knowledge grows, we redefine labels to better fit our understanding. At the present time, our knowledge about planets is expanding rapidly. We need to refine our definitions to fit what we know.

The Sun was once considered a planet; now we call it a star. The Moon was once identified as one of the seven planets; now we label it as a moon. The Earth was once the central object in the universe and most certainly not a planet; now the Earth is defined as a planet and not as the central object in the universe. Certainly, one thing we can learn from the history of astronomy is that we should not be afraid to recategorize objects when our knowledge of the universe changes substantially.

Pluto has been a planet for three-quarters of a century, but large numbers of astronomers and the public are ready to reclassify it as a large Kuiper Belt object and currently the largest (known) of the Plutinos. The basis for this desire to reclassify Pluto is entirely sociological; as we have seen, there is no reason to reclassify Pluto according to an old or new definition of *planet* that is based on physical criteria. Pluto is a planet.

Ceres was called a planet for at most a few decades, solely because it matched the Titius-Bode rule criterion. When the Titius-Bode rule failed, the physical criterion for calling Ceres a planet disappeared; thus, for the last century and a half, Ceres has been called an asteroid. Yet Ceres is clearly different in a physically identifiable and meaningful way from a tiny object like Gaspra, and Ceres meets the definition of a planet. The time has come to call Ceres a planet.

At the least, it is clear that Ceres and Pluto, and perhaps a few more intermediate-sized objects in the solar system, including Vesta and Sedna, should be treated identically. Either they are all planets, or none of them are planets, or all of them are something else entirely. Are they *minor planets*? As currently used, *minor planets* is not a label that fits these objects because this term is used to include Sun-orbiting objects of any shape, including thousands that are far too small to be considered *planets*. As it is unlikely that we will cease to use the term *minor planets* for all small objects in orbit around the Sun, we should accept that this particular usage of the term *planet* is flawed. We should accept that the adjective *minor* when used to modify *planet* refers to objects that actually are not planets; such objects meet the first and third criteria for planets but are too small to meet the second criterion. Ceres, Vesta, Pluto, and Sedna should not be called minor planets. They are planets.

Reclassifying the Sun, the Moon, the Earth, and Ceres did not diminish our interest in those objects. In fact, our improved understanding of Pluto's place in the solar system should help us refocus and ask better questions about Pluto.

Scientifically, we are unable to define a planet in a sufficiently meaningful way such as to include Pluto without including many other objects that few, if any, astronomers or laypersons wish to label as planets; we also are unable to develop a definition based on principles of astronomy and physics that excludes Pluto in any nonarbitrary way. Clearly, Pluto meets all the requirements to be a planet, yet Pluto is much more than just a planet, in the same way that Ceres meets our requirements, but is much more than just a planet. Both are planets if it serves our purposes in learning about the universe and our place in the universe by calling them planets. *Both are planets if we stick to a definition based on physical principles. Both are planets.*

Our understanding of planet-sized objects in and beyond our solar system demands that we add additional classification levels in our descriptions of planets. Some planets may fit into more than one classification bin. To those who seek a one-size-fits-all solution, the whole list of adjectivally compromised planets—major planets, giant planets, orphaned planets, free-floating planets, pulsar planets, moon-planets, belt-embedded planets, gaseous planets, rocky planets, icy planets—will be unsatisfactory. But for those who seek a better understanding of the physical world, this zoo of different kinds of planets provides us with exactly the scientifically meaningful set of categories that will help us achieve a better understanding of the cosmos.

Our quest to answer the question *Is Pluto a planet?* led us directly to a question about physics: *What is a planet?* Answering this second question, which was not simple or easy, has revealed that we live in a solar system that is quite different from the one we thought we lived in: *The solar system has more than twenty planets! Ceres is both a planet and a member of the asteroid belt and therefore both a planet and an asteroid. Pluto is a planet and a Plutino and a KBO. 2003 UB$_{313}$ is a planet and an SDO. Earth is both a planet and a rocky planet. Jupiter is both a planet and a gas giant planet.*

If we are interested in questions about life in the universe, several moons (e.g., Europa, Titan, and Triton), planets (e.g., Ceres, Mars, and Pluto), and comets (e.g., Halley) may all be good places to look for answers; all of them might be hospitable to life of some kind. If we inquire, all of them may yield important evidence from which we may learn more about our place in the universe. All of them are valuable to study and understand.

If we are interested in memorizing lists with the names of the planets, in providing simple answers to young children, we should remember that simple answers are usually wrong and that young children, who always ask *why* after we provide them with simple answers, are extremely capable of learning about the wonders and nuances of our universe. For the youngest among us, the adjectives (giant, terrestrial, belt-embedded, icy, pulsar) will not be a problem. It is time to start teaching our youngsters something more complicated, with more depth of meaning, than the simple memorization rubric My Very Earthly Mother Just Served Us Nasty Pizza.

○
Postscript

Current Thoughts by Other Astronomers

The International Astronomical Union (IAU) (http://www.iau
.org/), founded in 1919, is the internationally recognized professional organization whose "mission is to promote and safeguard
the science of astronomy in all its aspects through international
cooperation." There are many other professional organizations
(e.g., the American Astronomical Society), but in the business of
naming and numbering planets, moons, asteroids, comets, and stars,
all other professional organizations defer to the wisdom and authority of the IAU.

Regarding the issue of Pluto and the definition of a planet, the
IAU issued the following statement (press release 01/99) in 1999:

> Recent news reports have given much attention to what was believed
> to be an initiative by the International Astronomical Union (IAU)
> to change the status of Pluto as the ninth planet in the solar system.
> Unfortunately, some of these reports have been based on incomplete
> or misleading information regarding the subject of the discussion and
> the decision making procedures of the Union.
>
> The IAU regrets that inaccurate reports appear to have caused
> widespread public concern, and issues the following corrections and
> clarifications:
>
> 1. No proposal to change the status of Pluto as the ninth planet
>   in the solar system has been made by any Division, Commission
>   or Working Group of the IAU responsible for solar system sci-
>   ence. Accordingly, no such initiative has been considered by
>   the Officers or Executive Committee, who set the policy of the
>   IAU itself.

2. Lately, a substantial number of smaller objects have been discovered in the outer solar system, beyond Neptune, with orbits and possibly other properties similar to those of Pluto. It has been proposed to assign Pluto a number in a technical catalogue or list of such Trans-Neptunian Objects (TNOs) so that observations and computations concerning these objects can be conveniently collated. This process was explicitly designed to not change Pluto's status as a planet.

Through the present time, the IAU continues to reiterate that its official position is unchanged:

> The Status of Pluto: See the IAU Press Release 01/99.
> NOTE: The Press Release expresses the position of the IAU regarding the status of Pluto. The IAU considers the discussion closed with this statement and does not intend to reopen it in the foreseeable future.

The IAU includes a Working Group on ExtraSolar Planets (WGESP) that is "charged with acting as a focal point for research on extrasolar planets." The WGESP has offered the following position statement on the definition of a planet, last modified by working group members Alan Boss, Paul Butler, William Hubbard, Philip Ianna, Martin Kürster, Jack Lissauer, Michel Mayor, Karen Meech, François Mignard, Alan Penny, Andreas Quirrenbach, Jill Tarter, and Alfred Vidal-Madjar in 2003:

> Rather than try to construct a detailed definition of a planet which is designed to cover all future possibilities, the WGESP has agreed to restrict itself to developing a working definition applicable to the cases where there already are claimed detections, e.g., the radial velocity surveys of companions to (mostly) solar-type stars, and the imaging surveys for free-floating objects in young star clusters. As new claims are made in the future, the WGESP will weigh their individual merits and circumstances, and will try to fit the new objects into the WGESP definition of a "planet," revising this definition as necessary. This is a gradualist approach with an evolving definition, guided by the observations that will decide all in the end.

Emphasizing again that this is only a working definition, subject to change as we learn more about the census of low-mass companions, the WGESP has agreed to the following statements:

1. Objects with true masses below the limiting mass for thermonuclear fusion of deuterium (currently calculated to be 13 Jupiter masses for objects of solar metallicity) that orbit stars or stellar remnants are "planets" (no matter how they formed). The minimum mass/size required for an extrasolar object to be considered a planet should be the same as that used in our Solar System.
2. Substellar objects with true masses above the limiting mass for thermonuclear fusion of deuterium are "brown dwarfs," no matter how they formed nor where they are located.
3. Free-floating objects in young star clusters with masses below the limiting mass for thermonuclear fusion of deuterium are not "planets," but are "sub-brown dwarfs" (or whatever name is most appropriate).

These statements are a compromise between definitions based purely on the deuterium-burning mass or on the formation mechanism, and as such do not fully satisfy anyone on the WGESP. However, the WGESP agrees that these statements constitute the basis for a reasonable working definition of a "planet" at this time. We can expect this definition to evolve as our knowledge improves.

Needless to say, the IAU position, in which they take no position, is unsatisfactory to many, whereas the WGESP position, in which they do take a position, satisfies no one. In fact, the WGESP point 1 states that the "*minimum* mass/size required for an extrasolar object to be considered a planet should be the same as that used in our Solar System" (emphasis added), yet it is exactly because we cannot agree on what this minimum mass is that we are debating the issue in the first place.

The debate over the status of Pluto includes many other knowledgeable voices.

Alan Stern and Harold F. Levison, both of the Southwest Research Institute in Boulder, Colorado, produced the following definition, in language intended for an audience of professionals. Note that they use the word "body" to exclude "distributed/ensemble astrophysical objects, such as molecular clouds, galaxies, etc., that are patently not planets" but that would otherwise meet this definition:

> A *planetary body* is defined as any body in space that satisfies the following testable upper and lower bound criteria on its mass: If isolated from external perturbations (i.e., dynamical and thermal), the body must:
>
> 1. Be low enough in mass that at no time (past or present) can it generate energy in its interior due to any self-sustaining nuclear fusion reaction (else it would be a brown dwarf or a star). And also,
> 2. Be large enough that its shape becomes determined primarily by gravity rather than mechanical strength or other factors (e.g., surface tension, rotation rate) in less than a Hubble time, so that the body would on this timescale or shorter reach a state of hydrostatic equilibrium in its interior.[1]

Stern and Levison note that their definition is quantitative, is based on observable characteristics, and does not depend on the location of the object. They then offer a set of classification types for planetary bodies. Some are based on dynamics: planets (single planetary bodies in orbit around a star or double-star system), planetary-scale satellites (planetary bodies in orbit around a larger planetary body), unbound planets (i.e., free-floaters). Others are based on mass: subdwarf planets (e.g., Pluto, Charon, Ceres, etc., with masses less than 3 percent of an Earth mass), dwarf planets (e.g., Ganymede, Mercury, Mars, Venus, Earth, with masses less than 10 Earth masses), subgiants (e.g., Uranus, Neptune, with masses less than 100 Earth masses), giants (e.g., Jupiter, with masses less than 1,000 Earth masses), and supergiants (none known, with masses less than 30,000 Earth masses). Others are based on composition: rock planets (Mercury, Venus, Earth, Mars, Ceres), hydrogen gas planets (Saturn, Jupiter), and ice planets (KBOs, Uranus, Neptune).

Marc W. Buie, an astronomer at Lowell Observatory who has led a team that has discovered hundreds of outer solar system objects, suggests the following definition:

1. An object must have a shape that is controlled by gravity, not material strength (usually that means spherical but tides and rotation can change this),
2. An object must not be and can never have been generating energy by any fusion reaction and must not contain degenerate matter (i.e., it is not now a stellar or brown dwarf remnant and has never been a star or brown dwarf).[2]

Buie suggests that all objects that meet these two criteria are planets, but that we should then apply meaningful modifiers to distinguish between different types of planets: primary planets (orbit stars), secondary planets (orbit planets), free planets (not bound to any specific object), terrestrial, icy, and gas planets, and so on. Jupiter, for example, would be a primary gas planet.

Gibor Basri, Professor of Astronomy at the University of California at Berkeley, solves the dilemma by defining three types of objects, based only on the physical characteristic of mass:

FUSOR: an object that achieves core fusion during its lifetime.
PLANEMO (shorthand for "planetary-mass object"): a round
    non-fusor.
PLANET: a planemo orbiting a fusor.[3]

Like Stern, Levison, and Buie, Basri argues for the liberal use of adjectives (minor planets, major planets, beltway planets, terrestrial planets, gas giant planets, ice dwarf planets).

Michael E. Brown, Associate Professor of Planetary Sciences at Caltech and the leader of the team that discovered Sedna, Quaoar, 2004 DW, 2005 FY$_9$, and most important, the super-Pluto-sized 2003 UB$_{313}$, among other outer solar system objects, has a very different view. At the time of his discovery of Sedna, Brown preferred a definition of *planet* that depended on a single criterion: a (presumably substellar) object orbiting a star is either a solitary individual or a member of a large population. The individuals (e.g., Earth, Venus, Uranus) are planets. The populations (e.g., objects in the asteroid belt or in the Kuiper Belt) are not. Pluto and Ceres and Sedna and Quaoar are members of populations and so are not planets. The other traditional eight planets are solitary objects (presumably ignoring any Trojan asteroids) and so are planets. Brown would call all spherical objects "planetoids," in which case all eight planets would be planetoids but not all planetoids would be planets.[4] However, after his discovery of 2003 UB$_{313}$, Brown announced that he and his team had discovered the tenth planet, despite the fact that 2003 UB$_{313}$ was clearly part of the extended KBO population and therefore, by his own criterion, is a planetoid but not a planet. He wrote:

> Some astronomers have rather desperately attempted to concoct solutions which keep Pluto a planet, but none of these are at all satisfactory, as they also require calling dozens of other objects planets. While people are perhaps prepared to go from 9 to 10 planets when something previously unknown is discovered, it seems unlikely that many people would be happy if astronomers suddenly said "we just decided, in fact, that there are 23 planets, and we decided to let you know right now." There is no good scientific way to keep Pluto a planet without doing serious disservice to the remainder of the solar system.[5]

He writes further, in defense of 2003 UB$_{313}$ being a planet but in contradiction to his earlier position on the KBO population:

> From now on, everyone should ignore the distracting debates of the scientists, and planets in our solar system should be defined not by some attempt at forcing a scientific definition on a thousands-of-years-old cultural term, but by simply embracing culture. Pluto is a planet because culture says it is.
>
> We are then left with two cultural choices. (1) Draw the line at Pluto and say there are no more planets; or (2) Draw the line at Pluto and say only things bigger are planets. Both would be culturally acceptable, but to me only the second makes sense. In addition, the second continues to allow the possibility that exploration will find a few more planets, which is a much more exciting prospect than that suggested by the first possibility. We don't think the number of planets found by the current generation of researchers will be large. Maybe one or two more. But we think that letting future generations still have a shot at planet-finding is nice.

Thus, we declare that the new object, with a size larger than Pluto, is indeed a planet. A cultural planet, a historical planet. I will not argue that it is a scientific planet, because there is no good scientific definition which fits our solar system and our culture, and I have decided to let culture win this one. We scientists can continue our debates, but I hope we are generally ignored.[6]

Brian G. Marsden, Associate Director for Planetary Sciences, Harvard-Smithsonian Center for Astrophysics, makes the point that the word *planet* by itself is not scientifically very useful:

> Though seemingly evoking a very basic astronomical concept, the ancient word "planet" may have been doomed as soon as it ceased to apply to the seven traditional sky wanderers and acknowledged instead the place of the earth in the Copernican revolution and Galileo's recognition of scarcely smaller objects themselves in orbit about Jupiter. Whether one is talking about a "terrestrial" or a "giant" planet in our solar system, or a "free-floating" planet in a young star cluster, it has rarely been scientifically useful to use the word without at least some qualification.[7]

Jack Lissauer, a planetary scientist at the NASA-Ames Research Center, argues for a fairly traditional interpretation, with mass as the primary criterion for planethood, though his position lacks a clear criterion for the lower mass boundary for planets:

> I believe that the definition of a planet should be based primarily if not exclusively on mass, with the boundary between planets and brown dwarfs being the mass at which an object fuses half of its deuterium (about 13 times the mass of Jupiter. . . . Stars are defined by their ability to sustain quasi-equilibrium fusion of ordinary hydrogen at some stage in their evolution, regardless of whether they orbit a more massive star or not. Analogously, brown dwarfs can be defined as objects massive enough to fuse a majority of their deuterium, but not sustain fusion of hydrogen. The term "planet" should apply only to bodies too low in mass for substantial fusion of any type. However, to correspond with the Solar System, very low-mass objects would not qualify as planets (they would be referred to as minor planets or aster-

oids), and objects orbiting a planet would be classified as moons rather than planets. Moons must be less massive than the planet they orbit . . . . , in most cases, it will be impossible to know the origin of a particular object with great confidence. Thus, we should avoid using origin hypotheses as part of the definition of "planet." However, in analogy to moons, which must orbit a planet, one might consider using "free-floating planetary mass objects" (and ultimately a new term) for small interstellar objects and reserve "planet" for objects orbiting a star.[8]

Michael F. A'Hearn, Professor of Astronomy at the University of Maryland, argues that Pluto, as a member of the Plutino group, is not a planet:

> Why do we, as scientists, care how Pluto (or anything else) is classi-fied? This question must be answered before we can intelligently deal with how to present the result of a classification to the public. Scien-tists put things into groups, the members of which share common properties, in order to find patterns that will enable us to better under-stand how the bodies work or how they became what they are. If we are interested in origins, then it is clear with our present understand-ing (which might change in the future) that free-floating bodies of mass comparable to Jupiter are not in the same category as Jupiter itself. Similarly, it is clear that Pluto is not a planet like Jupiter but rather a planet like the numerous Plutinos in the 3:2 libration with Neptune. Thus Pluto should be classified as the largest Plutino.[9]

George Wetherill, a planetary scientist in the Department of Terrestrial Magnetism at the Carnegie Institution of Washington in Washington, D.C., thinks Pluto must remain a planet, that the lower boundary limit for planets is arbitrary anyway, so we should just set that limit at Pluto:

1. Distinguish between a planet and a star by deuterium burning. Thus brown dwarfs are a kind of star. There will be borderline cases, but so what? Some day, when we understand formation of these bodies much better, finer distinctions can be made, but we don't know enough yet to make definitions based on theories of origin.

2. Don't downgrade Pluto, regardless of the reasons given by others. It can be called "the largest plutino" as well, by those who make use of this characteristic in their work, without removing its status as the smallest planet. How can that hurt anything? It doesn't imply that all planets formed in the same way. That's something we don't know yet and may not know for a long time.

3. There are just plain "planets," the size of Pluto or larger, and smaller bodies called "minor planets." The boundary is arbitrary, so why not just leave it where it is? The distinction between different kinds of minor planets (asteroid, comets, etc.) can also be made in a way that does not change traditional usages, but still be scientifically appropriate.

4. Planets (and minor planets as well) are defined by being gravitationally bound to one or more stars.

5. Bodies that would otherwise be classified as planets, but are not gravitationally bound to a star, can be called "free planets" (including free minor planets that certainly exist, even though they haven't been found yet). Again, this doesn't imply any assumptions of their origin.[10]

Nick Woolf, an astronomer at Steward Observatory at the University of Arizona, rejects the deuterium-burning threshold as an upper boundary and simply suggests arbitrary mass boundaries as ten times bigger than Jupiter and ten times smaller than Mercury (which would eliminate Pluto as a planet):

The current reality is that we have no observations that suggest that any system is substantially like our own. And the word "planet" has developed a suite of meanings historically connected with our own system, so that to give it to other potentially quite different objects is guaranteed to produce confusion. Indeed, as our methods of finding objects of low mass improve, we are also going to be trying to match occultation observations, gravitational lens observations, astrometric observations and imaging observations.

I suggest that while the confusion rages, and until a paradigm seems to be gathering consensus, we give all these objects that we consider to have a likely close relationship to planets a single name—"PMOs," planetary mass objects, pronounced "peemose." This name would be restricted to objects which seem likely to be no more than an order of magnitude more massive than Jupiter, and no less than an order of magnitude below Mercury. Clearly there will be some fringe objects, where we are uncertain what the mass is, and these can be called probable or possible PMOs.

I too reject the deuterium burning limit as being an appropriate boundary. It seems unlikely to play any significant role in the formation or evolution of these objects. Certainly, when we get a better model that explains the likely histories of these objects and systems, we may find that we will reclassify a few objects. But the nomenclature proposed does allow us a substantial amount of breathing space until we converge on a paradigm.[11]

Dale Cruikshank, a research scientist with NASA and president of the IAU's commission for the physical study of the planets and satellites in the solar system, walks the line by letting Pluto be everything to everyone:

> The commission's official position is that Pluto is one of the nine planets in the solar system, a status that the faraway object has held since it was discovered in 1930.
>
> My personal view is that Pluto should probably have dual citizenship, in that its planetary status ought to be maintained partly for historical reasons and partly for its physical characteristics. But it seems clear that it is also "object one" in what we now recognize as a large class of Kuiper Belt objects.
>
> If Pluto does get that designation, either for common use or eventually by some official body such as the IAU, then it should also retain its status as a planet.[12]

◯
Appendix

What We Know about Pluto

DISTANCE FROM THE SUN

Pluto is a spherical object composed of rock and ice, but mostly ice, located on average almost 6 billion km (4 billion miles) from the Sun. In comparison, Earth is a ball composed mostly of rock, located forty times closer to the Sun than Pluto, at an average distance of only 150 million km (93 million miles; 1 km is equal to 0.625—or 5/8—miles) from the Sun.

PHYSICAL SIZE

For astronomers, sizes and distances are so big that we find it most useful to express most of what we know about other astrophysical objects in relative terms. Pluto's diameter (2,300 to 2,400 km) is 0.19 (or 19 percent) times the diameter of the Earth, which is 12,756 km at the equator. For comparison, Jupiter is the biggest planet in orbit around the Sun, with a diameter of 142,984 km (11.2 Earth diameters). Saturn (120,536 km), Uranus (51,118 km), and Neptune (49,552 km) also are much bigger than the Earth. On the other size side, Venus (12,104 km, or 0.95 Earth diameters), Mars (6,787 km, or 0.53 Earth diameters), and Mercury (4,878 km, or 0.38 Earth diameters) all are smaller than the Earth. Pluto is less than half as big as Mercury, the second smallest of the nine traditional planets. In fact, seven moons are bigger than Pluto: the Earth's moon (3,476

Figure A.1: Image of Pluto (*lower left*) and Charon (*upper right*) obtained with the Hubble Space Telescope. (Image courtesy of NASA/JPL-Caltech)

km), Jupiter's moons Io (3,630 km), Europa (3,130 km), Ganymede (5,280 km), and Callisto (4,840 km), Saturn's moon Titan (5,150 km), and Neptune's moon Triton (2,700 km).

SHAPE

Normally, we think of the Earth as round, by which we mean spherical. It is not, however, shaped like a perfect sphere, as the diameter from pole to pole is 12,712 km, 44 km (0.3 percent) less than the equatorial diameter. If the shape of the Earth were determined entirely by the force of gravity, it would be perfectly spherical, but two factors contribute to the slight deviations of the Earth from perfect sphericity. The first of these, the mechanical strength of the crust, permits the presence of little bumps and dips—mountains and valleys—that are scattered across the face of the Earth. The little bumps and dips found all over the surface are very small (the largest is less than one-tenth of 1 percent of the radius of the Earth) compared to the diameter of the Earth and are fairly uniformly distributed over the surface: the peak of Mount Everest stretches to a height just under 9 km above sea level, while the Marianas trench in the Pacific Ocean plunges almost, but not quite, 11 km below sea level. Thus, these effects are too small to cause us to think of the Earth as non-

233

Figure A.2: Hubble Space Telescope image of Saturn, illustrating the oblateness of this fast-spinning ball of gas. (Image courtesy of NASA/JPL-Caltech)

spherical. The second factor, the fast rotation of the planet combined with the softness of the warm inside of the Earth, spins the equatorial regions outward, making the Earth an *oblate spheroid*, fatter from one side of the equator to the other than from North Pole to South Pole.

All the other traditional planets, from tiny Pluto to giant Jupiter, also are spherical. Of all the planets, Jupiter and Saturn, being rapidly spinning balls of gas, deviate the most from perfect sphericity, having polar diameters fully 6 percent and 10 percent smaller than their equatorial diameters. Uranus (2.3 percent), Neptune (1.7 percent) and Mars (0.6 percent) are more perfectly spherical than Jupiter and Saturn. Both Venus and Mercury rotate so slowly that for each planet, the polar and equatorial diameters are identical (as best as we can measure them), so they are very nearly perfect spheres. Pluto is so far from Earth that we cannot measure the polar and equatorial diameters to an accuracy of 1 percent; thus, to within the limitations of our ability to make measurements of Pluto, it is as perfectly spherical as Venus and Mercury. Pluto also rotates very slowly, as we will soon discuss, and so we predict that we will find it to be essentially perfectly spherical, once we are able to make the necessary measurements.

MASS AND DENSITY

Since all of these objects are very nearly spherical and the volume of a sphere is proportional to the cube of its diameter, we find that the volume of Pluto is about 150 times less than that of the Earth and 8.5 times less than that of Mercury. Therefore, if the Earth and Pluto were made of the same kind of material, we would expect Earth to be about 150 times more massive than Pluto; yet, the Earth's mass, the total amount of stuff out of which it is made, is 476 times greater than that of Pluto (the mass of Pluto is 0.0021 times the mass of the Earth). This comparison immediately tells us that the mass density of Pluto is very different from the Earth's.

The mass density is a number that is calculated directly as the ratio between the known mass and volume of an object. In turn, the mass density immediately gives us information about the materials that compose the bulk of the planet. Objects with densities greater than that of water sink, while those with lesser densities float in water. For planets, good rough numbers for use in comparing densities include typical densities of about 1,000 kg per cubic meter for water (or water ice), about 3,000 kg per cubic meter for rock, and about 5,500 kg per cubic meter for iron. A volume of helium or hydrogen gas would have much lower densities than water or water ice.

At the surface of the Earth, we often think of a 1 kg object as equivalent to a 2.2 lb. object; however, kilograms and pounds are different kinds of things. At the surface of the moon, where the downward gravitational pull is much less than at the surface of the Earth, this same 1 kg object would now have a weight that is six times smaller than it had on Earth, barely 0.35 lb. Thus, we should think of mass (measured in kg) as "amount of stuff" and weight (which we like to measure in units of pounds) as "how hard gravity is pulling" on that mass.

From the mass and volume of Pluto, we can immediately determine its density, 1,750 kg per cubic meter. This density is less than one-third (32 percent) that of the Earth. The difference in density between these two planets tells us that the bulk of Pluto is made of different kinds of materials than the Earth. Pluto is composed probably of about 70 percent hydrated rock and 30 percent water ice, while the Earth is mostly rock and iron. Other than Pluto, the least massive planet is Mercury, which is eighteen times less massive than the Earth. Mercury, however, has a higher average density than the Earth and thus appears to be made almost entirely of iron.

ORBIT

Orbits are characterized by several parameters that define, mathematically, the shape and size of an orbit. The most important of these are the *semi-major axis* and the *eccentricity*.

The *semi-major axis* (*a*) is one-half of the long axis (the *major axis*) of an ellipse. The semi-major axis of the Earth's orbit is defined as one *astronomical unit* and is abbreviated as AU. One AU is equal to 149,600,000 km (93,000,000 miles). When someone asks how far the Earth is from the Sun, the answer typically given is, 1 AU; however, the Earth is rarely exactly 1 AU from the Sun. Instead, the distance of the Earth from the Sun varies continually throughout the year.

The *eccentricity* (*e*) of an orbit is a measure of the deviation from circularity of the ellipse. A perfectly circular orbit has an eccentricity of 0.00, while an object whose orbit is open, like a horseshoe, has an eccentricity of 1.00 or greater. A few comets have eccentricities greater than 1.00; these objects, after passing near the Sun, speed away from the Sun into interstellar space, never to return. The eccentricity of the Earth's orbit is 0.0167, very close to but not exactly zero. The very small value of the eccentricity tells us that the shape of the Earth's orbit is a good approximation to a circle; however, the orbit of the Earth is not a circle with the Sun at the center. The Sun is off-center and the Earth-Sun distance changes every day. The closest approach of the Earth to the Sun, a position called *perihelion*, is a distance of about 147 million km; the most distant point in the orbit of the Earth around the Sun, the position termed *aphelion*, is not much greater, at approximately 152 million km. These distances are found by multiplying the semi-major axis of the Earth's orbit (149.6 million km) times the quantity $1 - e$ (1.0000 − 0.0167) for perihelion and times $1 + e$ (1.0000 + 0.0167) for aphelion. Thus, at its most distant point in its orbit, the Earth is only 3 percent farther from the Sun than at the closest point in its orbit. Note that the Earth is at a distance of exactly 1 AU from the Sun for only two brief moments during each orbit, as it moves from perihelion to aphelion and back again.

Now we can discuss the technical aspects of Pluto's orbit. Like Mercury, Pluto's orbit is very elliptical (or, as astronomers prefer to say, eccentric), whereas the orbits of Earth, Venus, Mars, Jupiter, Saturn, Uranus, and Neptune all are much more circular. Pluto's eccentricity is 0.249, almost fifteen times larger than the eccentricity of the orbit of Earth. The orbital eccentricity of Mercury is fairly similar to that of Pluto, at 0.206, with Mars in third place with 0.093.

As Pluto travels through its elliptical orbit, it is sometimes much closer to the Sun and sometimes much farther from the Sun than its characteristic semi-major axis distance of 39.53 AU. Pluto's aphelion distance is 49.37 AU, but its perihelion distance is only 29.69 AU. Consequently, Pluto's distance to the Sun changes by more than 60 percent.

SOLAR INSOLATION CHANGES AND TEMPERATURE

The amount of light received by an object varies inversely with the square of the distance from the source of light. Consequently, Pluto receives 2.77 times more sunlight at perihelion than aphelion.* In turn, if the Sun were the only factor that controlled the temperature of a planet, then the temperature would be proportional to the amount of received sunlight to the one-fourth power (this is called the Stephan-Boltzmann law); thus, since $2.77^{1/4} = 1.3$, Pluto would

* $2.77 = (49.37 / 29.69)^2 = (1.66)^2$.

be 30 percent warmer at perihelion than at aphelion, all other things being equal. Naturally, all other things are not equal. On the Earth and Venus, greenhouse gases in the atmosphere have an enormous effect on the actual surface temperature of the planet. On Pluto, the ways in which Pluto's thin atmosphere may moderate the seasonal effects of changes in solar insolation remain unknown.

Measurements indicate that when Pluto was near perihelion, it had surface regions with temperatures as low as 40 K (−390° F) and other regions with temperatures as high as 60 K (−350° F).

PLUTO'S SURFACE AND ATMOSPHERE

Spectra of Pluto reveal evidence of several kinds of frost—carbon monoxide (CO), nitrogen (N_2), and methane (CH_4)—on the surface. All three of these molecules also are thought to be components of Pluto's atmosphere. Nitrogen is presumed to be the most abundant atmospheric constituent, but parts of the surface may be nitrogen frost–free. Nitrogen also is assumed to be the dominant component of the atmosphere. The surface pressure could be as high as fifty microbars (twenty thousand times less pressure than at the surface of the Earth) but may be as low as three microbars. Pluto's atmosphere is likely capable of slowly escaping from Pluto because of Pluto's weak gravitational grip on atmospheric gases.

Pluto has an atmosphere, perhaps for only a few decades of every orbit, perhaps always. For at least a few decades of every orbit, when Pluto is closest to the Sun, the surface of Pluto is warm enough that some nitrogen and methane and carbon monoxide frosts can sublimate—they change phases directly from solids to gases without becoming liquids—and form an atmosphere. Some planetary scientists predict, however, that for most of Pluto's orbit, these gases will freeze out as surface frosts, in the same way that water freezes out of our own atmosphere and, indeed, out of Mars's atmosphere on cold winter mornings; other planetary scientists believe, however, that Pluto may retain an atmosphere throughout its orbit. Mercury is worse off than Pluto. Mercury has no permanent atmosphere at all, only a haze of potassium and sodium atoms stripped off its surface that rapidly escape Mercury's weak gravity into space. But the other planets all have significant atmospheres. Even Titan, Saturn's largest moon, has an atmosphere so thick that visible light cannot penetrate through the atmosphere to Titan's surface.

CLOSER OR MORE DISTANT THAN NEPTUNE

Neptune, generally recognized as the eighth planet from the Sun, has a semi-major axis of 30.06 AU and an almost perfectly circular orbit, with an eccentricity of 0.009. Thus, Neptune is never farther than 30.33 AU from the Sun and never closer than 29.79 AU to the Sun. Consequently, for 16 years of each 248-year orbit, when Pluto is approaching and passing perihelion, Pluto is closer to the Sun than Neptune. Pluto was closer to the Sun than Neptune from 1979 until 1999; on the eleventh of February 1999, at 11:22 UT (6:22 a.m. EST), Pluto again became more distant than Neptune. Pluto will next cross Neptune's orbit on the fifth of April 2231.

ROTATION

Like all other known objects in the universe, Pluto rotates on its own axis. Amazingly, despite the enormous distance to Pluto, in the early 1950s, Merle F. Walker and Robert Hardie were able to measure the rotation period of faint Pluto. In 1965, Hardie refined their first measurement of 6.390 ± 0.003 days to 6.3867 ± 0.0003 days (the error on this measurement indicates that we know the rotation period of Pluto to an accuracy of about twenty-six seconds).

In our solar system, most of the planets stand close to straight up; that is, their rotation axes are almost perpendicular to their orbital planes. For the Earth, the rotation axis is tilted 23.5 degrees from the ecliptic plane (this tilt is the primary cause of the seasons);

Jupiter's tilt is only 3.1 degrees. Pluto's tilt is 122.5 degrees. This enormous tilt means that Pluto is closer to upside down than to right side up. Alternatively, we can think of Pluto as being right side up but spinning *backward.*

MOONS

Like the Earth and all of the planets except Mercury and Venus, Pluto has at least one moon. Pluto's largest moon, Charon, discovered in June 1978 by U.S. Naval Observatory astronomers James Christy and Robert Harrington, is named after the boatman from Greek mythology who ferried dead souls across the river Styx into Hades, the domain of Pluto. Charon has a diameter of 1,200–1,300 km, almost exactly half that of Pluto itself. This gives Charon a volume and mass about one-eighth that of Pluto, making the Pluto-Charon system the closest thing in our solar system to a binary planet. The next most massive moon in comparison to its planet is our own Moon, for which the mass ratio of the Moon to Earth is 1:81. Charon has a surface covered in water ice but no detectable atmosphere.

Charon orbits Pluto in 6.387 days at a distance of only 19,600 km. The orbital period (the time needed for one object to go completely around another object) of Charon is identical to its own rotation period and also to the rotation period of Pluto. In comparison, our Moon orbits in 27.32 days at a distance of 384,000 km; however, whereas our Moon also has a rotation period exactly equal to its orbital period, the Earth's rotation period is 1 day, not 27.32 days.

For a moon, having identical revolution and rotation periods— a situation referred to as synchronous rotation—is normal and is the consequence of the influence of tides raised on a moon by its planet and on the planet by its moon(s); this seemingly unusual situation is true not just for the Pluto-Charon and Earth-Moon systems, but also for most of the moons of most of the planets in the solar system. Pluto, however, is the only planet whose rotation has been slowed by the tides all the way to synchronicity. In comparison, tides raised on the Earth by the Moon have slowed the rotation period of the Earth from a 5-hour day, early in the history of the Earth, to the modern 24-hour day; but the Earth will never slow down all the way to a 656-hour day (i.e., a rotation period equal to 27.32 of our current 24-hour days).

On October 31, 2005, Hal Weaver and Alan Stern, on behalf of a large team, reported the discovery of two small moons of Pluto, found in images of Pluto taken by the Hubble Space Telescope on May 15 and 18, 2005. The moons, which were first discovered on June 15 in an inspection of the Hubble Space Telescope images by team member Max Mutchler and were then given the preliminary designation S/2005 P 1 (the brighter one) and S/2005 P 2, are both about 5,000 times fainter than Pluto. In June 2006, the IAU approved the names Hydra for S/2005 P 1 and Nix for S/2005 P 2. In mythology, Hydra is a nine-headed monster with the body of a serpent and Nix (using Egyptian rather than Greek spelling) is the goddess of darkness and night. Both moons appear to have nearly circular orbits, in Pluto's equatorial plane, as does Charon, with orbital periods of 38.2 ± 0.8 days (Hydra) and 25.5 ± 0.5 days (Nix). These orbital periods appear to be very close to six and four times the orbital period of Charon. Hydra probably has a diameter in the range of 110–160 km while Nix is likely 10 to 15 percent smaller. The masses of both small moons are likely only a few ten-thousandths of the mass of Pluto.

O
Notes

Chapter 1: What Is a Planet?

1. *Oxford English Dictionary*, http://dictionary.oed.com/, copyright © Oxford University Press 2003.

Chapter 2: Seven Perfect Planets
Made of Aether

1. *Lamp at Midnight* in *Cardozo Studies in Law and Literature* 2 (2) (Fall–Winter 1990): 1–6 to 1–8.
2. *Science in the Middle Ages*, David C. Lindberg, University of Chicago Press (Chicago), 1978, p. 279.

Chapter 3: The Earth Becomes a Planet

1. *Science in the Middle Ages*, David C. Lindberg, p. 286.
2. *The Book of the Cosmos*, ed. Dennis R. Danielson, Helix Books (Cambridge, MA), 2000, p. 92.
3. Adapted, with comments added, from *Cosmology: Historical, Literary, Philosophical, Religious, and Scientific Perspectives*, ed. Norriss S. Hetherington, Garland Publishing (New York), 1993, p. 152.
4. This usage of *world* is comparable to our modern usage of *universe*. To Copernicus, the world encompasses everything that exists, everywhere.
5. *The Copernican Revolution*, Thomas S. Kuhn, Harvard University Press (Cambridge, MA), 1957, p. 190.
6. Olaf Pedersen, "Galileo and the Council of Trent: The Galileo Affair Revisited," *Journal for the History of Astronomy* 14 (February 1983): 15.
7. *The Galileo Affair*, Maurice A. Finocchiaro, University of California Press (Berkeley), 1989, p. 335.
8. The Copernican Revolution, Thomas S. Kuhn, p. 191.
9. Ibid., p. 191.
10. As quoted in *The Book Nobody Read*, by Owen Gingerich, Walker & Company (New York), 2004, p. 138.

Chapter 4: Sixteen Planets

1. *Sidereus Nuncius*, Galileo Galilei, trans. Albert van Helden, University of Chicago Press (Chicago), 1989, p. 26.

2. *The Extraterrestrial Life Debate, 1750–1900*, Michael J. Crowe, Dover Publications (Mineola, NY), 1999, p. 11.

3. *Cardinal Bellarmine's Certificate of 26 May, 1616*, as quoted in *The Galileo Affair*, by Maurice A. Finocchiaro, p. 153.

4. *Sentence of 22 June, 1633*, as quoted in *The Galileo Affair*, by Maurice A. Finocchiaro, p. 287ff.

5. *A History of Astronomy*, A. Pannekoek, Dover Publications (Mineola, NY), 1989, p. 255.

6. No other satellites of Saturn would be discovered until the end of the eighteenth century (by William Herschel) or of Mars (in 1877 by Asaph Hall) or of Jupiter (in 1892 by E. E. Barnard) until the end of the nineteenth century; and despite great effort, no satellites have been discovered around Mercury or Venus.

7. *The Extraterrestrial Life Debate*, Michael J. Crowe, p. 19.

8. *The Book of the Cosmos*, ed. Dennis R. Danielson, p. 212–14.

9. *Plurality of Worlds*, Steven J. Dick, Cambridge University Press (Cambridge), 1982, p. 131.

Chapter 5: Not Everything That Orbits the Sun Is a Planet

1. *The Comet Is Coming*, Nigel Calder, Viking Press (New York), 1980, pp. 17–18.

2. *The Origin of Comets*. M. E. Bailey, S.V.M. Clube, and W. M. Napier, Pergamon Press (Oxford), 1990, pp. 50–55.

3. Ibid., p. 10.

4. As quoted in *Never at Rest: A Biography of Isaac Newton*, by Richard S. Westfall, Cambridge University Press (Cambridge), 1980, p. 403.

5. As quoted in *The Comet Is Coming*, by Nigel Calder, p. 40.

Chapter 6: Uranus!

1. According to Bode, he first found this idea in a translation by Titius of the book *Observations Concerning Nature*, by Bonnet. Bode notes that the idea was not found in the original edition of the book, only in the translation in a footnote added by the editor (*A Source Book in Astronomy*, Harlow Shapley and Helen Howarth, McGraw-Hill Books [New York], 1929, p. 181).

2. First part: *Out of the Darkness the Planet Pluto*, Clyde Tombaugh and Patrick Moore, Stackpole Books (Harrisburg, PA), 1980, p. 46.

3. Second part: *Planets X and Pluto*, William Graves Hoyt, University of Arizona Press (Tucson), 1980, p. 25.

4. Any number raised to the zeroth power (as would be true in this equation when $N = 2$) is equal to one.

5. *The Extraterrestrial Life Debate*, Michael J. Crowe, p. 76.

6. *The Herschel Chronicle: The Life-Story of William Herschel and His Sister, Caroline Herschel*, Constance A. Lubbock, Cambridge University Press (Cambridge), 1933, p. 59, as quoted in *William Herschel and the Construction of the Heavens*, Michael A. Hoskin, Oldbourne Book Co. (London), 1963, p. 19.

7. *A History of Astronomy*, A. Pannekoek, p. 312.

8. Ibid., p. 313.

Chapter 7: The Celestial Police

1. Osservatorio Astronomico di Palermo Giuseppe s. Vaiana, "Bode's Law and the Discovery of Ceres," by Michael Hoskin, http://www .astropa.unipa.it/versione_inglese/Hystory/BODE'S_LAW.htm

2. As quoted by Bode, in *A Source Book in Astronomy*, Harlow Shapley and Helen Howarth, p. 182.

3. The *eccentricity* of an orbit is a measure of the deviation from circularity of the ellipse. The eccentricity can be as small as 0.00 for a perfectly circular orbit; larger values of the eccentricity, which range from 0 to 1 for elliptical orbits, indicate increasingly flattened, or elongated, elliptical orbits.

Chapter 8: Neptune, the Thirteenth Planet

1. *Planets X and Pluto*, William Graves Hoyt, p. 33.

2. Ibid., p. 34.

3. Ibid., p. 37.

4. Ibid., p. 38.

5. The nineteenth-century usage of the term *Planet X* referred to the unknown next planet astronomers expected to discover, not to the "tenth" planet, which is how this designation is sometimes understood in modern usage.

6. Encke himself would write that "the prompt discovery was possible only because of the excellent Academy Star Charts by Bremiker; the disk can be recognized only when one knows it exists" (letter of September 26, 1846, printed in *Astronomische Nachrichten* 580 [October 12, 1846], as

quoted in *A Source Book in Astronomy*, Harlow Shapley and Helen Howarth, p. 254).

7. One (the Moon) for the Earth; four (Io, Europa, Ganymede, Callisto) for Jupiter; eight (Titan, Iapetus, Rhea, Tethys, Dione, Mimas, Enceladus, Hyperion) for Saturn; one (Triton) for Neptune.

Chapter 9: Easy Come, Easy Go

1. *Watchers of the Skies*, Willy Ley, Viking Press (New York), 1963, p. 417.

2. *Planets X and Pluto*, William Graves Hoyt, p. 71.

3. *Elements of Astronomy*, Charles A. Young, Ginn & Company (Boston), 1901, p. 241.

4. *In Search of Planet Vulcan*, Richard Baum and William Sheehan, Plenum Trade (New York), 1997.

5. *Elements of Astronomy*, J. Norman Lockyer, D. Appleton & Company (New York), 1881, p. 73.

6. *Elements of Astronomy*, Charles A. Young, p. 241.

Chapter 10: Pluto, the Fourth Ninth Planet

1. *Planets X and Pluto*, William Graves Hoyt, p. 77.

2. Ibid., p. 77.

3. Ibid., p. 77.

4. Ibid., p. 78.

5. Ibid., p. 80.

6. As quoted in "The Prediction and Discoveries of Pluto and Charon," by Gibson Reaves, in *Pluto and Charon*, ed. S. Alan Stern and David J. Tholen, University of Arizona Press (Tucson), 1997, p. 11.

7. From Lowell Observatory Observation Circular, "The Discovery of a Solar System Body Apparently Trans-Neptunian," V. M. Slipher, March 13, 1930, Flagstaff, AZ.

8. *Out of the Darkness the Planet Pluto*, Clyde Tombaugh and Patrick Moore, p. 151.

9. E. M. Standish Jr., "Planet X: No Dynamical Evidence in the Optical Observations," *Astronomical Journal* 105 (1993): 2000–2006.

10. Ibid.

Chapter 11: The Hidden Secrets of the Outer Solar System

1. Asteroid designations begin with the year of discovery followed by two letters. The first letter indicates the half month of discovery, beginning

with *A* for the first half of January (letter *I* is omitted and letter *Z* is unused). The second letter indicates the chronological order of discovery in that half month, with the first discovered object being designated *A* and the twenty-fifth designated *Z* (again, letter *I* is omitted). If more than twenty-five objects are discovered in one half month period, the second letter is recycled and includes the subscript 1, with the numerical subscript incrementing by one at object numbers 26, 51, 76, and so on. Thus, 1977 UB was the second object reported during the second half of October 1977.

2. *Beyond Pluto*, John Davies, Cambridge University Press (Cambridge), 2001, p. 25ff.

3. Ibid., p. 14.

4. Ibid., p. 41.

5. Brian G. Marsden, IAUC (International Astronomical Union Circular) 5730, March 29, 1993.

6. See the *Kuiper Belt Electronic Newsletter: Objects & Observations: Published Observations*, http://www.boulder.swri.edu/ekonews/objects/tabobs .html, for an updated list.

Chapter 12: The Plutinos

1. *Pluto and Charon*, Alan Stern and Jacqueline Mitton, Wiley & Sons (New York), 1998, p. 154.

Chapter 13: Is Pluto a Planet?

1. *Planets X and Pluto*, William Graves Hoyt, p. 207.

2. David Jewitt, "Pluto," http://www.ifa.hawaii.edu/faculty/jewitt/kb/ pluto.html (accessed December 27, 2005).

3. "Is Pluto a Giant Comet?" http://cfa-www.harvard.edu/cfa/ps/icq/ ICQPluto.html (accessed December 27, 2005).

Chapter 14: Goldilocks

1. From Einstein's famous equation $E = mc^2$. E is the energy, m is the mass, and c is the speed of light. One helium nucleus has less mass than four protons; in the fusion process, the difference in mass between the four protons and the one helium nucleus is the mass m that is released as energy.

2. An up-to-date list of Trojans is maintained by the Minor Planet Center at the Harvard-Smithsonian Center for Astrophysics: "Trojan Minor Planets," http://cfa-www.harvard.edu/iau/lists/Trojans.html (accessed December 27, 2005).

3. For a complete, up-to-date list, see the *Extrasolar Planets Encyclopaedia*, http://vo.obspm.fr/exoplanetes/encyclo/catalog.php, copyright 2005, Jean Schneider, updated December 24, 2005.

4. From "Definition of a Planet," http://exoplanets.org/science.html, November 6, 2000.

5. The IAU Minor Planet Center, at http://cfa-www.harvard.edu/iau/mpc.html, tracks the orbits of asteroids, comets, and Kuiper Belt objects.

Postscript: Current Thoughts by Other Astronomers

1. S. A. Stern and H. F. Levison, "Regarding the Criteria for Planethood and Proposed Classification Schemes," in *Highlights of Astronomy*, vol. 12, ed. H. Rickman, Astronomical Society of the Pacific (San Francisco), 2002, pp. 2005–13.

2. Personal communication.

3. Gibor B. Basri, "What Is a 'Planet'?" *Mercury*, November/December 2003, pp. 27–34.

4. Michael Brown, "Sedna," http://www.gps.caltech.edu/~mbrown/sedna/.

5. Michael Brown, "The Discovery of 2003 UB313, the 10th Planet," http://www.gps.caltech.edu/~mbrown/planetlila/.

6. Ibid.

7. Brian G. Marsden, "Dual Classification for Pluto and Other Bodies," http://www.sciencemag.org/cgi/eletters/291/5508/1487b?ck=nck#268, copyright © 2001, American Association for the Advancement of Science, February 21, 2001.

8. Jack Lissauer, "What Is a 'Planet'?" http://www.sciencemag.org/cgi/eletters/291/5508/1487b?ck=nck#267, copyright © 2001, American Association for the Advancement of Science, February 22, 2001.

9. Michael F. A'Hearn, "Classification Finds Patterns," http://www.sciencemag.org/cgi/eletters/291/5508/1487b?ck=nck#273; copyright © 2001, American Association for the Advancement of Science, March 1, 2001.

10. George Wetherill, "Common Usage Is Compatible with Scientific Value," http://www.sciencemag.org/cgi/eletters/291/5508/1487b?ck=nck#276, copyright © 2001, American Association for the Advancement of Science, March 22, 2001.

11. Nick Woolf, "Black Sheep and Pre-paradigm names," http://www.sciencemag.org/cgi/eletters/291/5508/1487b?ck=nck#273, copyright © 2001, American Association for the Advancement of Science, April 2, 2001.

12. In "Dual Citizenship Seen for Pluto: Join the Debate!" Leonard David, http://www.space.com/scienceastronomy/solarsystem/plutos_citizenship_010130.html, January 30, 2001.

Index